Affinity Separations

The Practical Approach Series

SERIES EDITORS

D. RICKWOOD
Department of Bio Biology, University of Essex, Wivenhoe Park, Colchester, Essex CO4 3SQ, UK

B. D. HAMES
Department of Biochemistry and Molecular Biology University of Leeds, Leeds LS2 9JT, UK

★ indicates new and forthcoming titles

Affinity Chromatography
★ Affinity Separations
Anaerobic Microbiology
Animal Cell Culture (2nd edition)
Animal Virus Pathogenesis
Antibodies I and II
★ Antibody Engineering
★ Applied Microbial Physiology
Basic Cell Culture
Behavioural Neuroscience
Biochemical Toxicology
Bioenergetics
Biological Data Analysis
Biological Membranes
Biomechanics—Materials
Biomechanics—Structures and Systems
Biosensors
★ Calcium-PI signalling
Carbohydrate Analysis (2nd edition)
Cell–Cell Interactions
The Cell Cycle
Cell Growth and Apoptosis
Cellular Calcium
Cellular Interactions in Development
Cellular Neurobiology
Clinical Immunology
★ Complement
Crystallization of Nucleic Acids and Proteins
Cytokines (2nd edition)
The Cytoskeleton
Diagnostic Molecular Pathology I and II
Directed Mutagenesis
★ DNA and Protein Sequence Analysis
DNA Cloning 1: Core Techniques (2nd edition)
DNA Cloning 2: Expression Systems (2nd edition)
★ DNA Cloning 3: Complex Genomes (2nd edition)

★ DNA Cloning 4: Mammalian Systems (2nd edition)

Electron Microscopy in Biology

Electron Microscopy in Molecular Biology

Electrophysiology

Enzyme Assays

★ Epithelial Cell Culture

Essential Developmental Biology

Essential Molecular Biology I and II

Experimental Neuroanatomy

★ Extracellular Matrix

Flow Cytometry (2nd edition)

★ Free Radicals

Gas Chromatography

Gel Electrophoresis of Nucleic Acids (2nd edition)

Gel Electrophoresis of Proteins (2nd edition)

Gene Probes 1 and 2

Gene Targeting

Gene Transcription

Glycobiology

Growth Factors

Haemopoiesis

Histocompatibility Testing

HIV Volumes 1 and 2

Human Cytogenetics I and II (2nd edition)

Human Genetic Disease Analysis

★ Immunochemistry 1

★ Immunochemistry 2

Immunocytochemistry

In Situ Hybridization

Iodinated Density Gradient Media

★ Ion Channels

Lipid Analysis

Lipid Modification of Proteins

Lipoprotein Analysis

Liposomes

Mammalian Cell Biotechnology

Medical Bacteriology

Medical Mycology

Medical Parasitology

Medical Virology

★ MHC 1

★ MHC 2

Microcomputers in Biology

Molecular Genetic Analysis of Populations

Molecular Genetics of Yeast

Molecular Imaging in Neuroscience

Molecular Neurobiology

Molecular Plant Pathology I and II

Molecular Virology

Monitoring Neuronal Activity

Mutagenicity Testing

★ Neural Cell Culture

Neural Transplantation

★ Neurochemistry (2nd edition)

Neuronal Cell Lines

NMR of Biological Macromolecules

Non-isotopic Methods in Molecular Biology

Nucleic Acid Hybridization

Nucleic Acid and Protein Sequence Analysis
Oligonucleotides and Analogues
Oligonucleotide Synthesis
PCR 1
PCR 2
Peptide Antigens
Photosynthesis: Energy Transduction
Plant Cell Biology
Plant Cell Culture (2nd edition)
Plant Molecular Biology
Plasmids (2nd edition)
★ Platelets
Pollination Ecology
Postimplantation Mammalian Embryos
Preparative Centrifugation
Prostaglandins and Related Substances
Protein Blotting
Protein Engineering
★ Protein Function (2nd edition)
Protein Phosphorylation
Protein Purification Applications
Protein Purification Methods
Protein Sequencing
★ Protein Structure (2nd edition)
★ Protein Structure Prediction
Protein Targeting
Proteolytic Enzymes
Pulsed Field Gel Electrophoresis
Radioisotopes in Biology
Receptor Biochemistry
Receptor–Ligand Interactions
RNA Processing I and II
★ Subcellular Fractionation
Signal Transduction
Solid Phase Peptide Synthesis
Transcription Factors
Transcription and Translation
Tumour Immunobiology
Virology
Yeast

Affinity Separations

A Practical Approach

Edited by

PAUL MATEJTSCHUK
Research and Development Department,
Bio Products Laboratory, Elstree, Herts WD6 3BX, UK

Oxford University Press, Walton Street, Oxford OX2 6DP

Oxford New York
Athens Auckland Bangkok Bogota Bombay Buenos Aires
Calcutta Cape Town Dar es Salaam Delhi Florence Hong Kong
Istanbul Karachi Kuala Lumpur Madras Madrid Melbourne
Mexico City Nairobi Paris Singapore Taipei Tokyo Toronto

and associated companies in
Berlin Ibadan

Oxford is a trade mark of Oxford University Press

Published in the United States
by Oxford University Press Inc., New York

A catalogue record for this book is available from the British Library

Library of Congress Cataloging in Publication Data
(Data available)

ISBN 0 19 963551 X (Hbk)
ISBN 0 19 963550 1 (Pbk)

Typeset by Footnote Graphics, Warminster, Wilts
Printed in Great Britain by Information Press, Ltd, Eynsham, Oxon.

Preface

There can be few techniques in biological sciences that have found so wide a field of application as affinity chromatography. From molecular biology, through protein chemistry, and even in the realm of applied cell biology, affinity-based separations are commonplace. In producing a methods book on affinity separation my aim has been to provide straightforward and practical examples which have been proven in the laboratory together with sufficient discussion and background to allow the reader to adapt the principles described for use with their own application. This is not a textbook of theory nor is it intended primarily to satisfy the experienced chromatographer. Rather it is aimed at those beginning their laboratory careers or those who wish to develop an affinity purification scheme rapidly.

The authorship has been drawn from a wide range of specialities. In choosing chemical engineers to describe the recent developments in the principles of chromatography, the importance of the nature of the matrix' conjugation chemistry and ligand choice in achieving the desired purifications is stressed. For those seeking quantitative data from their affinity methods a chapter devoted to this topic has been included. Areas in which affinity separations have attained an unrivalled position amongst other forms of chromatography include the use of engineered affinity tags and also the purification of antibodies, so much so that I have devoted individual chapters to each of these topics. In order to capture the breadth of application, chapters are committed to cell separation technology and carbohydrate analysis (focusing especially on lectin affinity methods). The variety of affinity separation strategies available for the isolation of proteins and nucleic acids is considered in two specific chapters.

I would sincerely like to thank all the authors who have contributed. Thanks also go to David Hames, the series editor, and to the staff of Oxford University Press for their advice and support. I am grateful to my colleagues at Bio Products Laboratory, especially Drs John Davis and Mike Harvey, for constructive comments which have helped with the editing of this book. I hope that it will provide a useful text for those involved in the practice of affinity-based purifications and those seeking to explore this area for the first time.

Elstree P. M.
April 1997

Contents

4. Affinity separations of nucleic acids 99

Ram P. Singhal and Chenna R. Chakka

Contributors

CHENNA R.CHAKKA
Department of Chemistry, Wichita State University, Wichita, KS 67260-0051, USA.

RICHARD D. CUMMINGS
University of Oklahoma Health Sciences Center, Department of Biochemistry & Molecular Biology, BSEB 329, 941 S. L. Young Boulevard, Oklahoma City, OK 73190, USA. Fax: (01)-405-271-3910. email: richard-cummings@uokhsc.edc

PETER A. FELDMAN
Research & Development Department, Bio Products Laboratory, Dagger Lane, Elstree, Herts WD6 3BX, UK.

MIGUEL A. J. GODFREY
Clifmar Associates, School of Biological Sciences, University of Surrey, Guildford, Surrey GU2 5XH, UK. Present address: Veterinary Medicines Directorate, Woodham Lane, New Haw, Addlestone, Surrey KT15 3NB, UK. Fax: (0)1932-336618. email: m.godfrey@umd.maff.gov.uk

ERIK LUNDGREN
Department of Cell & Molecular Biology, University of Umeå, Umeå, S90187, Sweden. Fax: (0)46-90-111420.

ANDREW LYDDIATT
School of Chemical Engineering, University of Birmingham, Edgbaston, Birmingham, B15 2TT, UK. Fax: (0)121-414-5324.

PAUL MATEJTSCHUK
Research & Development Department, Bio Products Laboratory, Dagger Lane, Elstree, Herts WD6 3BX, UK. Fax: (0)181-207-5290. email: pm20489@bpl.co.uk

SUDESH B. MOHAN
School of Chemical Engineering, University of Birmingham, Edgbaston, Birmingham B15 2TT, UK.

JOHN MORE
Research & Development Department, Bio Products Laboratory, Dagger Lane, Elstree, Herts WD6 3BX, UK.

JACKY ROTT
Research & Development Department, Bio Products Laboratory, Dagger Lane, Elstree, Herts WD6 3BX, UK.

SATISH K. SHARMA
Biochemistry, Pharmacia and Upjohn 7240-267-117, 301 Henrietta Street, Kalamazoo, MI 49007, USA. Fax: (01) 616-833-1488. email: satish.k.sharma@am.pnu.com.

RAM P. SINGHAL
Department of Chemistry, Wichita State University, Wichita, KS 67260-0051, USA. Fax: (01)-316-689-3431. email: rpsingha@wsuhub.uc.twsu.edu

DONALD J. WINZOR
Centre for Protein Structure, Function & Engineering, Department of Biochemistry, University of Queensland, Brisbane, Queensland 4072, Australia. Fax: (0)61-73365-4699. email: winzor@biosci.uq.edu.au

Abbreviations

AAT	α_1-antitrypsin
aa-tRNA	amino-acyl transfer RNA
AEB	*N*-[*N'*-(*m*-dihydroxyborylphenyl)succinamyl]aminoethyl
3aPBA	3-aminophenylboronic acid
cAMP	cyclic adenosine monophosphate
Cb	clenbuterol
CD	cluster differentiation
CDI	carbonyl diimidazole
CDP	cytosine diphosphate
CE	capillary electrophoresis
CGE	capillary gel electrophoresis
CHT	ceramic hydroxyapatite
CMB	*N*-(*m*-dihydroxylboryl-phenyl)carbonylmethyl cellulose
CNBr	cyanogen bromide
Con A	concanavalin A
CV	column volume
CZE	capillary zone electrophoresis
DADPA	diaminodipropylamine
DEAE	diethyl aminoethyl
DMSO	dimethyl sulfoxide
dsDNA	double-stranded DNA
DTT	dithiothreitol
DVS	divinyl sulfone
EACA	ε-amino caproic acid.
EDC	1-ethyl 3-(3-dimethylaminopropyl)-carbodiimide
EDTA	ethylenediaminetetraacetic acid
EGTA	ethylenebis(oxyethylenenitrilo)tetraacetic acid
ELISA	enzyme-linked immunosorbent assay
FACS	fluorescence-activated cell sorter
FMP	2-fluoro-1-methyl-pyridinium toluene-4-sulfonate
FPLC	Fast Protein Liquid Chromatography
Fuc	fucose
γ-IFN	gamma-interferon
Gal	galactose
Gal*N*Ac	*N*-acetyl galactosamine
GC-MS	gas chromatography–mass spectrometry
GlcNAc	*N*-acetyl glucosamine
GST	glutathione S-transferase
GuHCl	guanidinium chloride
Hepes	2-hydroxyethyl-1-piperazine ethane sulfonic acid

HIC	hydrophobic interaction chromatography
HIV-1	human immunodeficiency virus, type 1
HPIAC	high performance immunoaffinity chromatography
HP(L)AC	high performance liquid affinity chromatography
HPLC	high performance liquid chromatography
IAC	immunoaffinity chromatography
ICAM-1	intracellular adhesion molecule 1
IDA	iminodiacetic acid
IgA	immunoglobulin A
IgG	immunoglobulin G
IgM	immunoglobulin M
IgY	immunoglobulin Y
IL-2	interleukin-2
IL-2R	interleukin-2 receptor
IL-4	interleukin-4
IMAC	immobilized metal affinity chromatography
ITP	isotachophoresis
LFA-1	lymphocyte function-associated antigen-1
mAb	monoclonal antibody
MAFF	Ministry of Agriculture Food and Fisheries
Man	mannose
MBE	moving boundary electrophoresis
MEKC	micellar electrokinetic capillary chromatography
MeOH	methanol
Mes	2-*N*-morpholinoethanesulfonic acid
Mops	3-(*N*-morpholino)-propane sulfonic acid
m.p.	melting point
NCAM	neuronal cell adhesion molecule
NeuAc	*N*-acetyl-neuraminic acid
NGF	nerve growth factor
NHS	*N*-hydroxysuccinimide
NMR	nuclear magnetic resonance
NTA	nitrilotriacetic acid
OHTb	hydroxytrenbolone
pAb	polyclonal antibody
PBS	phosphate-buffered saline
PEG	polyethylene glycol
PMSF	phenylmethyl sulfonyl fluoride
PNGase F	protein:N-glycanase F
QAE	quaternary aminoethyl
rIL-2	recombinant interleukin-2
rIgG	rabbit immunoglobulin
RNase	ribonuclease
RPB	reversed phase boronate

rProA	recombinant Protein A
RT	reverse transcriptase
SDS–PAGE	sodium dodecyl sulfate–polyacrylamide gel electrophoresis
SLAC	serial lectin affinity chromatography
SP	sulfopropyl
SpA	staphylococcal Protein A
SpA-PS	staphylococcal Protein A–porous silica
SPE	solid phase extraction
ssDNA	single-stranded DNA
StpG	streptococcal Protein G
StpH	streptococcal Protein H
TAC	thiophilic adsorption chromatography
TBS	Tris-buffered saline
TC	tresyl chloride (2,2,2, trifluoroethane sulfonyl chloride)
TEA	triethylamine
TNF	tumour necrosis factor
tPA	tissue-type plasminogen activator
VKD	vitamin K- dependent
ZE	zonal electrophoresis

1

Recent developments in affinity separation technologies

SUDESH B. MOHAN and ANDREW LYDDIATT

1. Introduction

Since the innovative work of Porath *et al.* (1) in the 1960s, and the seminal texts of Lowe and Dean (2) and Dean *et al.* (3), the technologies of affinity separation have become widely accepted—perhaps now to the extent of being taken for granted. Practitioners range widely from laboratories interested in the isolation of bioactive probes and tools for biological research, to large manufacturing plants active in the production of diagnostic and therapeutic agents.

During the last decade affinity chromatography applications have increased steadily in number and with them the number of affinity adsorbents available for laboratory-scale purifications. However, despite the obvious advantages of affinity chromatography, the technique has been slow to catch on for process-scale purifications and has only recently begun to find favour for the purification of high-value products. This is probably due to the inherent cost of immobilizing ligands on supports as well as to the problems posed by ligand leakage, the complicated validation procedures, and the need for adsorbents which can withstand the sanitation regimes. The ligands used conventionally tend to be large molecules with specific binding sites for the product and, unless careful consideration is given to the coupling conditions and the orientation of the bound ligand, the immobilization step often either renders them inactive or reduces their binding capacity. The recent developments of pseudoaffinity chromatography where the ligand is non-biological (as, e.g., in immobilized dye, metal, and thiol affinity chromatography) have increased interest in affinity chromatography because these ligands are inexpensive, chemically stable, and readily coupled to a variety of support matrices, thus reducing the potential cost of the overall purification methods. Such advances have been paralleled by the development of consumable products and supporting hardware by a variety of manufacturing companies. These have enabled the custom assembly of purification systems which can be empirically tailored to approach individual requirements. However, the technical literature which supports such products is often compromised by commercial considerations and affiliations. Hence, a text such as this volume is potentially invaluable.

Most end-users of the technology are principally interested in the successful recovery of sufficient product as the starting point for innovative experiments more than in the mechanics of the isolation process itself. This has encouraged a view of affinity separations cast as black art, and has severely limited a full exploitation of the technology through the natural transition of highly selective separations from the research laboratory to the manufacturing plant. In particular, the requirements of the regulatory authorities (4) in respect of the detailed description of manufacturing operations, operating protocols, and product qualities have exposed our limited mechanistic understanding of many biospecific phenomena which are readily exploited for research purposes in the laboratory.

This chapter gives an overview of affinity chromatography and deals with recent developments in the context of the factors that affect the efficacy of the technique. Details of some protocols, relevant to the issues discussed, are included. However, the reader is well advised to consult an earlier book in this series, *Affinity chromatography: a practical approach* (3), for numerous protocols which still remain useful.

2. Affinity separations—a generalized view

Figure 1 summarizes the key features of an affinity separation of a bioactive molecular product. In the first instance, the product will commonly be sourced in a complex mixture of micro- and macromolecular solutes arising from a fermentation broth, a cell homogenate, or some other form of biological extract. For simplicity in this chapter, we will mostly consider protein products, but the principles discussed herein are broadly applicable to all biomolecules. Most protein products have discrete biological activities or functions (e.g. as an enzyme, antibody, or hormone) which are commonly dependent upon recognition of unique molecular features (e.g. active sites, antigenic epitopes, or receptor binding domains) in dynamic interactions with other biomolecules (e.g. enzyme inhibitors and cofactors, immunogens, or receptors) which possess equally unique structural features. Immobilization of one of such a pair of interactants as the ligand (e.g. an enzyme inhibitor) upon the surface of a stationary solid phase by means of a stable chemical coupling, which does not significantly diminish the biochemical specificity of that molecule, creates an assembly (*Figure 1b*) upon (and with) which selective association of the second interactant (e.g. the enzyme) may occur.

When such a product is applied in a complex mixture of bioactive solutes (impurities) to an affinity matrix, a specific association occurs at their expense and is maintained under washing procedures. These manipulations are commonly undertaken in buffered salt solutions (*Figure 1c*, I and II) to exclude impurities further. Limited perturbation of the specific molecular

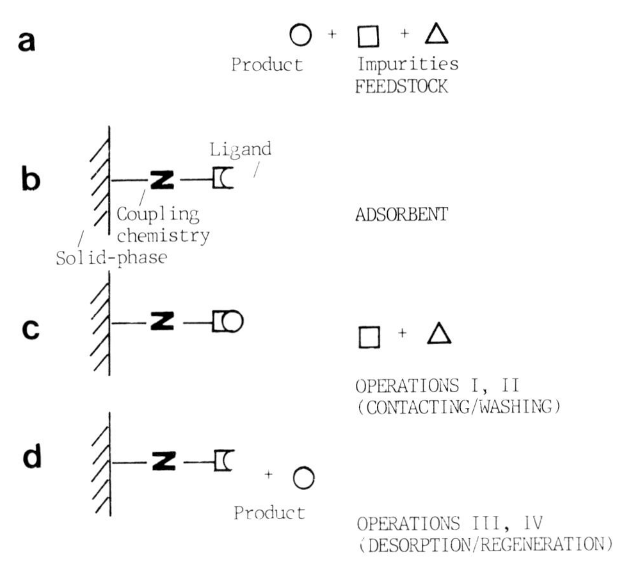

Figure 1. Schematic diagram of the main components and stages of a typical affinity separation. Feedstock (a) comprising product and impurities is brought into contact with a solid phase (b) derivatized with specific ligands. Product is selectively adsorbed at the expense of impurities (c) and desorbed (d) by application of appropriate mobile phase. The solid phase is finally regenerated for further use.

interaction between the product and immobilized ligand by manipulation of the liquid mobile phase (pH, ionic strength, solute content, or temperature) promotes dissociation of the product and facilitates potentially quantitative recoveries at high purity in the mobile phase (*Figure 1d*, III). Replacement of the desorption conditions, together with any necessary cleaning and re-equilibration (*Figure 1d*, IV), prepares the affinity assembly for a further cycle of adsorptive application.

Figure 1 clearly indicates that the nature of the coupling chemistry will profoundly affect the efficiency of an affinity separation process. The 'packing density' of ligands on the solid-phase surface will affect working capacities up to a point, beyond which steric interference (particularly between macromolecular ligands and products) may diminish performances. The molecular length of the coupling chemistry (generally expressed in carbon numbers (C_3, C_6, etc.) may promote a more efficient interaction of the ligand with a binding site buried within a three-dimensional domain of a protein product. Orientation of the ligand will also generally influence the efficiency of interaction with products. Simple ligands with only two interacting groups (i.e. with solid phase and ligand as in the protease inhibitor amino benzamidine, see ref. 5) have a predetermined molecular orientation offered for product interaction. Where the ligand is a protein (as in an antibody), commonly adopted methods of solid-phase coupling through surface amino groups necessarily impose a random orientation upon the immobilized ligand with an inevitable impact upon the global performance of any affinity adsorbent so constructed. Strategies for controlling ligand surface concentration and orientation are discussed later, but it should be emphasized that both will be affected by location within the internal geometry of the chosen solid phase.

Advances in our ability to quantify the kinetics of product adsorption and desorption (6), and to interpret such data in the context of ligand concentration and location (7) and the physical and biochemical conditions offered by the mobile phase behaviour (8), have greatly benefited the predictive design, assembly, and operation of affinity separation systems.

3. Support matrices used in affinity chromatography

Conventionally, the bioaffinity assembly depicted in *Figure 1* comprises the surface area of a porous, beaded particle which has a ligand molecule covalently and stably attached to it; the ligand guarantees continued selective association with a biospecific interactant (the product). The most efficient means of contact of such an assembly with feedstock solutions rich in that product has been a fixed bed contactor wherein specifically interacting products can be separated chromatographically from non-reacting impurities. Such operations gave rise to the term affinity chromatography, but a more accurate term might be affinity adsorption. In the best designed systems, only the product exhibits significant affinity for the immobilized ligand and thus the binding capacity of the adsorbent contactor is product-dedicated. This contrasts (see *Figure 2*) with non-specific separations such as ion-exchange adsorption where separation of products from impurities is based upon more subtle differences of affinity of solutes for the adsorbent at each theoretical stage or plate of the fixed bed separator (9). A practical consequence of bioaffinity adsorption is that operational procedures involving step changes of mobile phase conditions to progress between loading, washing, desorption, and re-equilibration stages are much simpler than those required to achieve fractionation on an ion-exchange adsorbent, which often requires gradient elution. This has positive implications for automation and process control in both the laboratory and production plant.

An extensive range of solid supports is now available for affinity chromatography. These are listed in *Table 1* in four groups although no attempt has been made to make this list exhaustive.

3.1 Particles as stationary solid phases

3.1.1 Supports for conventional affinity chromatography

The traditional and still the most popular matrix used for affinity chromatography is agarose (*Figure 3a*). Its large beads (40–160 μm in diameter) and

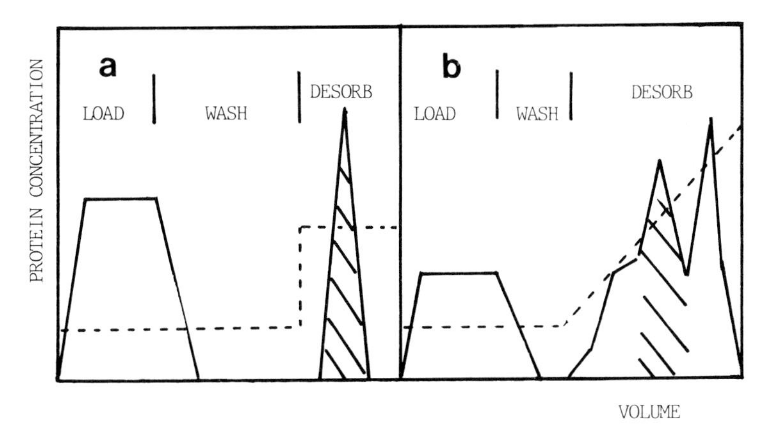

Figure 2. Comparison of the operation in fixed beds of (a) affinity-selective and (b) ion-exchange recovery of proteins from complex mixtures. Adsorbed product (hatched area) is (a) discretely desorbed by step changes (dashed line) of the mobile phase or (b) fractionally displaced by incremental modification of elution buffers (dashed line) to yield varying degrees of purification.

macroporous, accessible pore structure (100–300 nm pores) are well suited for use with large target molecules; for example, Sepharose CL-4B and CL-6B can accommodate molecules of 1 and 10 megadaltons, respectively. The advantages of agarose are high capacity, presence of functional groups (this eases matrix activation), good chemical stability especially at high pH (normally used to regenerate the solid phases), low non-specific binding, and good reproducibility. The drawbacks include its solubility in hot water (though cross-linking of agarose has now resulted in adsorbents which are stable to autoclaving) and non-aqueous solutions, susceptibility to microbial degradation, and, above all, a lack of rigidity which restricts its operation to low to medium pressure. Other gels introduced as alternatives to agarose have not been used widely although supports such as Prosep (Bioprocessing), Toyopearl (E. Merck), and Separon (Alltech Associates) might prove to be the matrices of choice in future.

3.1.2 Small, rigid beads for high-performance liquid affinity chromatography

To counteract the lack of rigidity of soft gels, numerous inorganic supports, based on silica and synthetic polymers, have become available. These supports are typically small porous beads (5–20 μm in diameter, with well-defined pore sizes; *Figure 3b*) and have the practical advantages of rigidity, strength, and reduced diffusion rates. These matrices can be operated in deep fixed beds at high linear velocities, using appropriate high pressure pumping systems, and hence have found applications mainly in high performance liquid affinity chromatography (HPLAC). Since the first introduction of

Table 1. Some of the particulate support materials available for affinity chromatography

Support material	**Trade name**	**Supplier**
Conventional affinity chromatography		
Agarose	Sepharose 2B, 4B, 6B; Sepharose CL-2B, CL-4B, CL-6B; Sepharose Fast Flow	Pharmacia
	Bio-gel A 50 m, 150 m	Bio-Rad
Agarose/polyacrylamide	Ultrogel	IBF (BioSepra)
Cross-linked dextran	Sephadex	Pharmacia
Dextran/polyacrylamide	Sephacryl	Pharmacia
Cross-linked cellulose	Matrex Cellufine	Amicon
Controlled pore glass	CPG	Pierce
	Prosep	Bioprocessing
Polyacrylamide	Bio-Gel-P	Bio-Rad
	Trisacryl	IBF (BioSepra)
	Eupergit C	Rohm Pharma
Methacrylate	TSK-Gel Toyopearl	E. Merck
	Separon	Alltech Associates
High performance liquid affinity chromatography		
Silica	Hypersil WP 300	Shandon
	Lichrospher	E. Merck
	Ultrasphere	Beckman
	Spherisorb	Phase Separations
	Zorbax	Rockland Technologies, Inc.
Methacrylate	Eupergit	Alltech Associates
Polymer	Affiprep	Bio-Rad
New materials		
Polystyrene	Poros-50	PerSeptive Biosystems
Polystyrene/hydrogel	HyperD	Sepracor (BioSepra)
Perfluorocarbons	–	Refs 12, 13
Zirconia particles	–	Refs 14, 15
Membranes as solid phases		
(a) Preactivated		
	Memtek	Amicon
Silica-PVC	Acti-Disk	FMC
Glass	Bioran-M	Schott Glass
(b) Ready to use		
	Trio membrane affinity system (protein A, protein G)	Sepracor
	MAC affinity membrane system (protein A, protein G)	Amicon
	Sartobind (dye ligands, metal chelate)	Sartorius

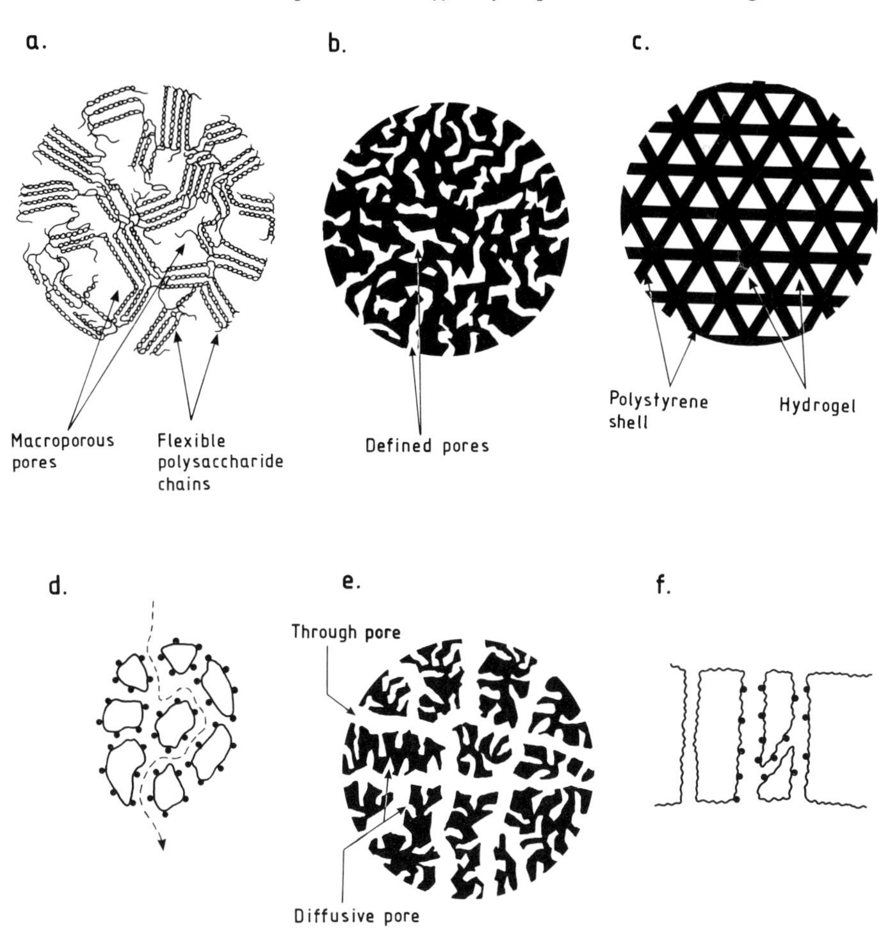

Figure 3. Schematic diagrams of the various types of solid phase particle. (a) An agarose-based particle with macroporous pores formed by the cross-linking of polysaccharide chains; (b) porous, rigid particle with defined pore sizes; (c), HyperD 'gel-in-a-shell' particle; (d) diagram to show ligand bound to the surface of the pores (solid circles) and its access to the product by diffusion (dotted line); (e) Poros particle containing through and diffused pores; (f) cross-section through a membrane with ligand (solid circles) immobilized in the pore channels.

HPLAC by Ohlson *et al.* (10) this technique has been used widely, but use has been restricted to analytical and laboratory-scale operations mainly because of the requirement for expensive high-pressure equipment. These supports are mainly based on silica though some polymer-based solid phases are also available. The main disadvantages of silica-based materials are non-specific adsorption due to the residual charge on the surface and solubility at alkaline pH. These effects can be minimized by extensive derivatization of the silanol groups on the surface of silica with functional groups: diol-bonded silica supports are available commercially and they can also be synthesized from a silane and porous silica as described in *Protocol 1*. Such functional groups shield the surface of the adsorbent from interaction with hydroxyl ions in the mobile phase and hence reduce the degradation at alkaline pH.

Protocol 1. Derivatization of silanol groups on silica adsorbents

Equipment and reagents

- Silica adsorbent
- Sodium-dried toluene
- Acetone
- Diethyl ether
- 3-Glycidoxypropyl trimethoxysilane (Aldrich)
- 0.05 M and 0.1 M HCl
- 3 M $Na_2S_2O_3$
- Methyl orange indicator
- Two-necked flask connected to a condenser for refluxing
- H_2SO_4-dried N_2
- Glass sintered funnel connected to a suction pump
- Oven to provide temperatures of 90 and 190°C

A. *Silanization of porous silica*

1. Dry silica (20 g) by heating at 190°C for 72 h in a 1 litre, two-necked flask.
2. Cool the reaction flask to room temperature. Add sodium-dried toluene (300 ml) followed by 3-glycidoxypropyl trimethoxysilane (20 ml). Stir the suspension gently to mix. Connect a condenser to one neck and use the second neck to blow a slow stream of H_2SO_4-dried N_2 over the surface. Reflux the mixture for 24 h at atmospheric pressure.
3. Wash the silane-coated particles (now bearing epoxy functional groups) on a glass filter with 5 volumes each of toluene, acetone, and ether; dry under vacuum if epoxy groups are to be used for further activations. Determine the epoxy group content as described in *Protocol 1B*.
4. The epoxy silica can be hydrolysed to diol silica by heating in 5 volumes of 0.05 M HCl at 90°C for 2 h.

B. *Determination of epoxy group content*

1. Mix 100 mg of silica with 1 ml of 3 M $Na_2S_2O_3$ in a test tube for 2–3 min to liberate hydroxyl ions from the epoxy groups. Dilute with 5 ml of distilled water.
2. Add methyl orange indicator and titrate against 0.1 M HCl.
3. Calculate the epoxy content from the quantity of acid used to neutralize the hydroxyl ions.

3.1.3 New supports

Two novel supports which appear to meet most of the criteria required for maximum productivity have recently come on the market. These are the HyperD and the Poros media, for HyperDiffusion and Perfusion chromatography, respectively.

HyperDiffusion chromatography, introduced by BioSepra, is based on HyperD beads (35 μm diameter) which comprise a polystyrene composite shell filled with a soft hydrogel (*Figure 3c*). This 'gel-in-a-shell' material combines the advantages of rigidity, due to the shell, and the high performance of the soft gels described above. It can withstand 2000 p.s.i. and can be operated at high linear velocities (> 5000 cm/h). Since the pores are filled with a homogenous hydrogel, binding sites are located throughout the pores rather than on the surface and hence the theoretical capacity of the support is high. The combination of high capacity and high throughput provides a significant improvement over the conventional supports. These beads are available as ready-to-use with Protein A, heparin, or Basilene Blue immobilized as ligands.

All the media listed above depend upon (and are limited by) diffusion and, though not seen as a practical problem in small-scale purifications, mass transfer across the adsorbent particles poses a problem at large scale (*Figure 3d*). The dynamic capacity of the adsorbent typically falls with increasing throughput and higher flow rates. This severely reduces the efficiency of the highly specific but highly costly ligands. The ligand is present on the surface of the support but in porous gels all the surface area is inside the particle. Hence if sufficient operational time is not allowed for diffusion, a significant proportion (10–30%, depending upon the system) of the ligand can remain unused. To circumvent this, PerSeptive Biosystems have introduced Poros polymeric particles, made of stable cross-linked polystyrene, for use in Perfusion chromatography (11). These particles contain two distinct classes of pores (*Figure 3e*): 'through pores' that penetrate the particle and are large enough to allow some convective flow through the particles, and smaller diffusive pores which line the through pores and provide a large adsorptive surface area. During a chromatographic run above a critical flow rate, convection dominates over diffusion in the through pores, allowing efficient access to the high surface-area diffusive pores of the particle. The combination of rapid perfusive transport within the through pores and the ultra-short (< 1 μm) diffusion path-lengths in the diffusive pores make both the resolution and capacity essentially independent of flow rate. The chemical and physical nature of the support is exceptionally robust. It is resistant to extremes of pH (1–14), and can withstand 0.5 M NaOH and 1.0 M HCl for cleaning and sterilizing purposes. Maximum pressure limit is 3000 p.s.i. which is higher than the HyperD support described above.

According to Afeyan *et al.* (11) the high throughput of perfusive particle columns allows great flexibility. Upon process applications, for example, a small column can be used in rapid cycles to produce a given throughput or amount of material in a specified time. These authors compared the effect of cycling mass throughput using soft gel, HPLC, and Perfusion systems and showed that Poros particles allowed the separations to be performed at 10–100 times the flow rate of HPLC and conventional techniques without loss in resolution or binding capacity. Typical run times for these techniques were 6 min, 1 h, and 4 h, respectively.

Poros material is available as 50 μm beads for low pressure operation or as smaller (20 μm) particles for analytical-scale HPLAC use. Matrices are available as aldehyde-, epoxy-, hydroxyl-, and hydrazide-activated beads as well as ready-to-use adsorbents with Protein A, Protein G, metal chelate, and heparin immobilized as ligands.

A third novel material, based on perfluorocarbons, has been described by Stewart *et al.* (12); it is not yet available commercially. Perfluorocarbons are chemically and biologically inert, high density (1.8–2.1 g/ml), and thermally stable synthetic materials which are totally insoluble in aqueous and organic solvents; their hydrophobic surface is wetted by water only in the presence of fluorosurfactants. For use as affinity adsorbent the authors coated the solid support with a fluorosurfactant (perfluorooctanoyl–polyvinyl alcohol) layer prior to coupling ligand. A further development of perfluorocarbons is their use as emulsions for use in fluidized beds. The first application of this methodology is the purification of human serum albumin from blood plasma (13).

Finally, modified porous zirconia has recently been added to the list of new emerging solid phases. Zirconia particles exhibit a high chemical stability in broad pH range (1–14) and have physical strength equal or superior to that of silica (14). Wirth and Hearn (15) have described the first applications of this material for affinity chromatography. These authors modified the surface of the particles by silanization and immobilized concanavalin A and iminodiacetic-Cu(II) for the adsorption of horseradish peroxidase and myoglobin, respectively.

3.1.4 Microporous membranes

The use of microporous membranes as support matrices is a recent idea (16). It involves immobilization of ligands on membranes (in the form of stacked sheets or hollow fibres) to provide an immense cross-sectional area but with short depth (100 μm; *Figure 3f*). Such systems approach the 'ideal' support in that the effects of diffusion are overcome and mass transfer limitations alleviated. In a packed-bed affinity column, mass transfer is limited by the time it takes for the solute molecules to diffuse through the interstitial spaces to the ligand itself which resides on the surface or within the pores of the particle. Assuming rapid binding kinetics the ligand would be used efficiently only if the average time for a solute molecule to diffuse to the ligand were substantially shorter than its residence time in the column. The development of the various solid phases described above has addressed this limitation by, e.g., reducing the particle diameter as in HPLAC matrices or design of particles for perfusion chromatography.

Affinity membranes operate in convective mode, which can significantly reduce diffusion limitations commonly encountered in column chromatography. As a result, a high throughput and faster processing times are possible in membrane systems. Membranes are also capable of handling unclarified solutions and thus can be applied in the earlier stages of downstream processing. Furthermore, since the productivity of an affinity adsorbent is dependent on ligand utility (see Section 4), the small depth of affinity membranes offers the added advantage of making efficient use of expensive ligands by maximizing their utilization. However, practical problems are posed by the design of feedstock contactors having sufficient capacity for the product. Some are stacked membrane sheets or discs packed into a contactor which mimics a fixed bed. Others have designs derived from the existing ultra- and microfiltration units which exploit hollow fibre, plate and frame, spiral wound, or pleated sheet configurations. All of these offer the short-diffusion distances of the sub-micrometre dimensions of membrane pore structures, but suffer from the 'dead volumes' required of retentate and permeate channels necessary to deliver feedstock to, and breakthrough away from, the bioactive surface. These will increase the size of the contactor (as compared with a fixed bed), and also diminish any fractionation and concentration of products achieved at desorption.

A number of affinity membranes are available commercially, either in pre-activated form or coupled to ligands such as Protein A (see *Table 1*). Methods for covalently attaching molecules to membrane materials are identical to those developed for solid beads. Activation and coupling reactions are carried out by recirculating solutions through the membranes and washing away any unreacted material. A series of recent publications have focused on non-chromatographic affinity systems using more robust and less expensive group specific ligands such as metal chelates and reactive dyes (17–19).

3.2 Alternative solid phases and contactors

Alternatives to the macroporous, beaded particles have been extensively examined, but the majority fail to provide sufficiently accessible and bioactive (in respect of specific binding) surface areas. Problems of providing sufficient surface area for bioaffinity association without limitations of diffusion resistances can be met by three further strategies which are currently the subject of intensive research. In the first the ligand is covalently attached to submicrometre diameter solid particles which provide sufficient surface area without the diffusion resistance of porous structures. Particle diameter limits the application of fixed-bed contactors, but suspension contacting in a batch reactor followed by enhanced particle recovery and manipulation for desorption appear as exciting possibilities. Perfluorocarbon emulsions (12,13) and magnetic solid particles (20) fall into this developing category of adsorbent system.

Secondly, gas bubbles stabilized with bifunctional surfactants and substituted with bioaffinity chemical groups, form colloidal gas aphrons (discussed in ref. 21) which could act in an analogous fashion to the solid particles discussed. Bubble generation, feedstock contacting and collection by flotation, and product desorption could prove to be an economic procedure at the largest scales.

A third strategy exploiting soluble polymers modified covalently by the addition of bioaffinity ligands has excited some interest. These polymer assemblies can be readily mixed in solution with a product-rich feedstock. Following product association, the polymers can be precipitated by modification of pH or ionic strength followed by product desorption (22). Alternatively, ligand-modified soluble polymers such as polyethylene glycol (PEG) can be introduced as a finely dispersed aqueous phase into a second aqueous phase (dextran- or phosphate-rich) containing the product feed (23). In common with all the contacting methods reviewed herein, the dispersed phase provides a large surface area for specific ligand association and is accompanied by selective phase transition by the target product. Subsequent phase separation by low-speed centrifugation generates a PEG-product top phase, and a bottom phase rich in impurities. The product may be readily back-extracted into a fresh bottom phase, and the affinity polymer assembly prepared for reuse.

3.3 Direct adsorption from biological suspensions: expanded/fluidized beds

All the solid phases and contactors discussed so far perform best with feedstocks clarified by centrifugation or microfiltration of suspended solids (cells, debris, and other particulates). Fixed bed contactors and some membrane devices will act as depth filters when challenged with any concentration of suspended solids. High performance (HPLC) solid phases (5–10 μm diameter) commonly require 0.2 μm-filtered feedstocks to minimize blinding even by dust and other fine particles. Alternatively, cross-flow membrane contactors will suffer fouling and resistance to product transport through gel-polarization phenomena. Only those newer developmental technologies outlined above which exploit solid particles, colloidal gas aphrons, and polymer precipitation or partition can potentially resist the problems encountered with suspended particles, although non-specific association/entrapment may still remain a problem requiring practical solution.

Solid–liquid separation of whole cells, cell debris, and other particulates from solutes in a biological feedstock is generally a mandatory requirement prior to the further fractionation of molecular mixtures and the subsequent recovery of a purified protein product. Complete clarification of feedstocks, particularly those sourced from cell homogenates which can be rich in submicrometre particles (fine debris, colloidal aggregates, membrane fragments, cell inclusions, etc.), can generally only be practically achieved by processes of ultracentrifugation. These suffer limitations of scale-up beyond the most modest of operations (> 0.1 litre) associated with process considerations of mechanical stress, material throughput, and economic cost (24). The circumvention of such problems by direct processing of biological suspensions without clarification has long been a goal of the process engineer, and in recent times has attracted the interest of researchers and commercial suppliers active in the development of processes of selective adsorption.

The processing of whole fermentation broths, cell homogenates, or other biological particulates through a liquid fluidized bed of selective adsorbent particles offers a strategy for the direct recovery of a product from 'difficult' feedstocks which conventionally require refined clarification by centrifugation or microfiltration. An input velocity can be chosen for the upward feed to a fixed bed of adsorbent particles such that the particles separate either into an expanded bed configuration (where particle mixing is limited, ref. 25) or at higher liquid velocities into a single stage, well-mixed fluidized bed (26; see *Figure 4a*). The nature of the expanded bed means that a finite number of theoretical plates facilitates a true chromatographic separation with a single pass of feedstock through the contactor. In contrast, the fluidized bed offers a single theoretical stage, which can approach an adsorption equilibrium between bound and unbound product by continuous recirculation of feedstock (*Figure 4b*). Alternatively, a multi-stage fluidized bed can be assembled wherein adsorbent moves against the flow of feedstock (*Figure 4c*; 27). Both expanded and fluidized bed strategies facilitate the use of suspension feedstocks without significant impact upon adsorption performance and subsequent product recovery. The latter is generally achieved (after thorough washing in fluidized bed mode) by flow reversal and fixed-bed desorption to facilitate maximal product fractionation, concentration, and purification. Clearly the use of appropriate bioaffinity adsorbents (as outlined above) would enable the direct, 'one-step' specific recovery of an intracellular product from an unclarified cell homogenate, or an extracellular product from a whole fermentation broth.

Bioselective expanded and fluidized bed operations require solid phases with properties additional to those associated with fixed-bed operations (see above). Principally, they must have material densities significantly higher than the basal agarose, polysaccharide, and other hydrophilic polymers currently used in the assembly of fixed-bed adsorbents. Values here generally approach that of the input feedstock which is close to that of water. Enhanced values of 1.2 g/ml for fluidized bed adsorbents are indicated, whilst higher values would enable greater fluid velocities, lower non-specific adsorption of debris and solute impurities, and shorter bed-residence times. Density

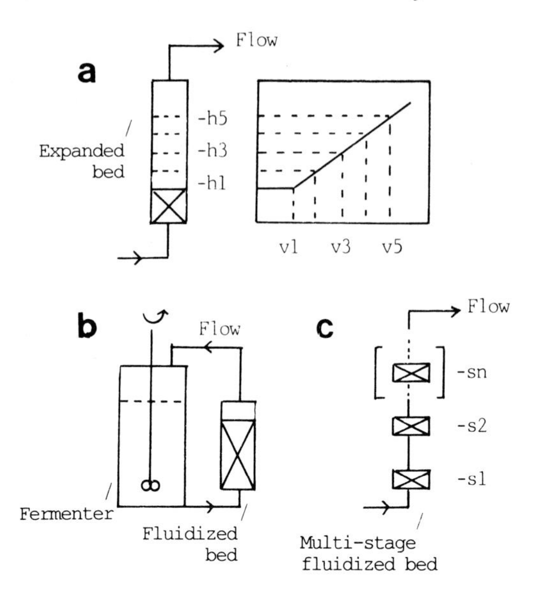

Figure 4. Representation of expanded and fluidized bed adsorbent contactors operated for the recovery of protein products from particulate feedstocks. In expanded beds (a), upward fluid velocities (v1–v5) yield bed heights (h1–h5) with minimal particle mixing and create multiple theoretical stages of adsorption suitable for single passes of the feedstock. Fluidized beds (b) are well mixed and offer a single theoretical stage for product adsorption in the recirculating loop of a fermenter. Multistage (s1–sn) fluidized beds (c) move counter-currently to a single pass feedstock. See Section 3.3 for details.

is currently enhanced by using composite particles assembled from biocompatible materials cast around, or mixed with, ceramic fillers. These latter clearly influence particle diameters (and diffusion resistances), limit the volumetric binding capacity of adsorbents, and may introduce additional problems of non-specific association of impurities and contamination of product streams through leakage and dissolution. Exposure of adsorbents during adsorption cycles to cells and debris, particularly in the extractive biotransformation outlined above, requires particular properties of physical and chemical resistance to subsequent cleaning, sanitization, and sterilization protocols associated with the operation and maintenance of homogeneous cell cultures. Although the criteria for the choice of chemistries of ligand coupling, and (to a degree) of the ligand itself, are similar for adsorbents proposed for both fixed and fluidized bed operations, the ideal solid phase suited for direct bioselective adsorption from whole broths or cell homogenates is not yet commercially available.

Various ion-exchange adsorbents have recently been marketed which perform well in direct adsorption from suspension feedstocks. Streamline (sulfopropyl (SP) and DEAE) from Pharmacia comprises cross-linked agarose beads (100–300 μm) cast around a quartz core to yield a particle density of 1.2 g/ml suitable for operation in commercial expanded bed devices (25). DEAE–Spherodex (a dextran–silica composite, 100–300 μm particles) from BioSepra has also proved effective in protein recovery in fluidized bed operations (28,29). Porous zirconia particles (14) have a high density and hence make another attractive potential solid phase for use in expanded and fluidized beds. Other ceramic–polysaccharide composites have been demonstrated in both ion-exchange and immunoaffinity recoveries of protein products from suspension feedstocks (29,30). It is clear that future wider availability of the basal solid phases of such adsorbents will enhance the development of materials for bioselective recoveries in fluidized or expanded bed contactors. Such materials will be derivatized by the controlled chemistries of activation and ligand coupling reviewed in the remaining sections of this chapter.

4. Performance of rigid solid phases

Unlike agarose, rigid matrices have closely defined pore sizes (*Figure 3*; (31,32)) and require a careful assessment of the solid phase used for a particular ligand/biomolecule system because resistance to access can affect the productivity of the solid phase. This has been documented by several authors and is illustrated here by the data reported by us (7,33). We studied the influence of pore size upon the productivity of three silica-based solid phases, Sorbsil C-200, C-500, and C-1000 (40–60 μm diameter particles with pore diameters of 20, 50, and 100 nm respectively; Crosfield Chemicals, Warrington, UK) and identified two unique parameters, biochemical productivity and maximum physical capacity, of the matrix as generically essential for the successful design and operation of productive affinity chromatography systems. Biochemical productivity, i.e. the molar ratio of the amount of the product recovered per unit volume of the adsorbent and ligand concentration, utilizes the expected stoichiometry of binding of the two components to assess the efficacy of the adsorbent. Maximum physical capacity, i.e. the percentage utilization of the theoretical maximum physical capacity of the matrix to accommodate the biomolecules, was calculated from the pore and the molecular dimensions assuming that there was no steric hindrance to access. Using a model system (immobilized human IgG–anti-human IgG monoclonal antibody system, in which both the ligand and the product are of 150 kDa), the amount of product obtained at different ligand densities was determined and from these values the maximum amount of ligand/product accommodated was calculated. It was shown that the physical capacity of C-200 was only 16% of the expected value. This capacity increased to 70 and 100% with C-500 and C-1000, respectively (*Table 2*). This was attributed to the distribution of the ligand, as detected by immunogold staining, within the silica particles (*Figure 5*): ligand was restricted to the peripheral 3 μm of the

Table 2. Characteristics of human-IgG matrices

Matrix	Human IgG challenge		*L* (mg/ml matrix)	*R* (mg/ml matrix)	*P* (*R*/*L*)	Physical capacity	
	mg/ml	Total mg/ml matrix				*R* + *L*	% expected value
C-200	1.3	3.0	2.1	0.6	0.3	2.7	15
	3.0	3.0	1.3*	1.13	0.9	2.4	13
	6.0	6.0	2.1	0.65	0.3	2.75	16
C-500	1.25	3.0	2.45	2.36	0.96	4.81	27
	2.5	3.0	2.9	5.8	2.0	8.7	48
	3.3	5.4	5.2	6.7	1.3	11.9	66
	3.3	7.5	7.2*	6.2	0.86	13.4	74
	1.0	9.0	5.7*	3.4	0.6	9.1	50
	9.0	9.0	7.4*	5.1	0.69	12.5	69
C-1000	2.5	4.5	4.0	7.3	1.8	11.3	63
	3.3	5.4	5.3*	9.7	1.8	15.0	83
	3.3	8.5	8.3	7.8	0.94	16.1	89
	1.0	9.0	9.0	9.1	1.0	18.1	101
	9.0	9.0	9.0	9.3	1.0	18.3	101

C-200, C-500, and C-1000 solid phases were challenged with varying concentrations of human IgG to obtain immunoadsorbents with a range of ligand concentrations (*L*). Fixed-bed characteristics were obtained by saturating 1 ml beds with an excess of purified anti-human IgG monoclonal antibody. Unbound antibody was removed by washing the beds with buffer and adsorbed product was eluted with 3 M KSCN and desalted immediately on Sephadex G-25 (1 × 10 ml). The amount of antibody recovered (*R*) was used to calculate the productivity (*P*, the molar ratio of *R*/*L*).
*The asterisks indicate the adsorbents examined for ligand location (Figure 5).

C-200 particles but was present throughout the C-1000 particles. In the case of C-500 the distribution of the ligand was dependent upon the initial ligand challenge used to prepare the adsorbent. Maximal productivity was achieved with an initial challenge of about 3 mg human IgG per millilitre: the ligand in this case was distributed uniformly throughout the particles, in contrast to adsorbents prepared with either low (1 mg/ml) or high (9 mg/ml) concentrations of human IgG. In both the latter cases the ligand was restricted to the periphery of the particles. Such dependency on the initial ligand concentration was not observed with the wider-pore-C-1000 adsorbent which exhibited the maximum (90–100%) expected capacity at equivalent ligand challenges (*Table 2*). Optimum productivity was achieved at the optimum ligand density, a concentration at which the bound ligand was utilized maximally.

The performance of an adsorbent also depends upon the size of the ligand/product (*Table 3*). Hence the capacity of C-200 was increased with a small ligand/biomolecule system (trypsin inhibitor/trypsin, 22.1 and 23.3 kDa, respectively): about 70% of the maximum physical capacity was available for interactions. Steric hindrance was demonstrated in the case of C-1000 using a large system (comprising immobilized monoclonal antibodies to β-galacto-

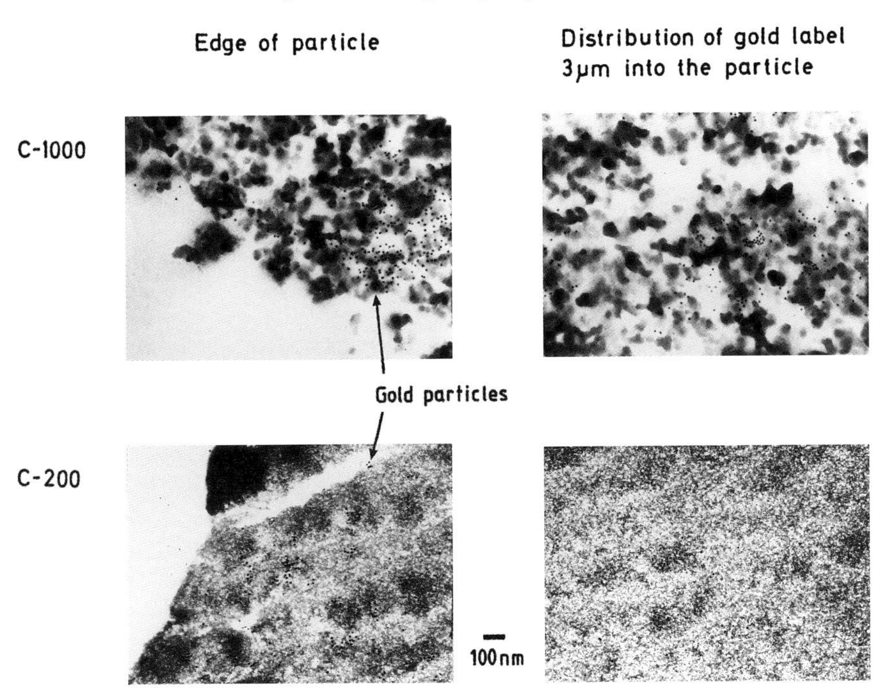

Figure 5. Immunogold-labelled electron micrographs of C-1000 and C-200 particles to show the distribution of ligand (human IgG) within these particles. Sections of the particles (10–15 nm thick) were incubated with immunogold conjugate (goat anti-human IgG, 10 nm gold; Biocell) and examined for gold label using a JOEL 100 CX temscan microscope. Micrographs on the left show the label detected at the edge of the particles. The distribution of label 3 μm into the particle is shown in the micrographs on the right. Only a few gold particles were detectable in C-200 (bottom right) relative to C-1000 particles (top right) suggesting restricted access in the former. Label was present throughout the C-1000 particles (not shown).

sidase, and β-galactosidase; 150 and 460 kDa, respectively), indicating that choice of support, appropriate to the size of the ligand/biomolecule being used, is necessary for maximum productivity.

5. Activation procedures

Activation procedures involve the modification of the surface of the solid phase to obtain functional groups to which the ligand can be coupled. These procedures generally result in a high density of these functional groups which require blocking after the reaction with the ligand is completed.

Agarose-based matrices activated with cyanogen bromide (CNBr) have historical significance and are still very popular. In this method cyanate esters and imidocarbonates are introduced by reacting the solid phases with CNBr

Table 3 Maximum capacities available for binding in three model systems

Model system	**% Capacity available for binding**		
	C-200	**C-500**	**C-1000**
Human IgG/anti-human monoclonal antibodies	16	70	100
Trypsin inhibitor/trypsin	70	nd	nd
Anti-β-galactosidase monoclonal antibodies/β-galactosidase	0	18	32

nd = not determined

at high pH and the ligand is coupled through an isourea linkage between the gel and the amino groups of lysine (34). CNBr is very toxic and produces toxic volatile gases during the activation procedure. However, the commercial availability of a stable activated gel in a lyophilized form has eliminated the need for handling toxic reagents and probably accounts for the extensive use of this procedure. The conditions used for binding the ligand are mild but result in an affinity adsorbent with undesirable ion-exchange properties as well as a problem of ligand leaching (see Section 6).

The newer activating chemistries have concentrated upon obtaining stable binding of the ligand with minimum residual charge. Numerous activation chemistries are now available to couple ligands through their amino, hydroxyl, carbonyl, and sulfhydryl groups. Of these, the commonly used ones include carbonyl diimidazole (CDI), epoxide, sulfonyl chlorides (tosyl and tresyl chloride), and *N*-hydroxysuccinimide. Carbonyl diimidazole produces an activated matrix which is relatively stable to hydrolysis but readily reacts with nucleophilic residues on protein molecules (35; *Figure 6*). Unreacted groups can be hydrolysed at pH 8.5–9.0 to release CO_2 and imidazole, reverting to the original hydroxyl groups of the matrix and hence leaving no residual groups for non-specific binding. Sulfonyl chloride activation was developed by Nilsson and Mosbach (36). Tresyl chloride appears to be the best activating agent for this procedure. Ngo (37) introduced 2-fluoro-1-methyl pyridinium toluene-4-sulfonate (FMP) to activate the hydroxyl groups of polysaccharide-based affinity adsorbents (*Figure 7*). The resulting linkages formed from reacting FMP-activated gel and amino- or thiol-containing ligands are stable as well as non-ionic. In addition, because 1-methyl-2-pyridine is released during the coupling reaction, the reaction can be followed conveniently by monitoring the absorbance at 297 nm using the molar coefficient of 5900. Furthermore, FMP-activated gels can be used to couple ligands containing either sulfhydryl or amine groups and provides a method which guide the ligand molecule to a particular orientation on the matrix (see *Protocols 3* and *6*). *N*-hydroxysuccinimide esters, which are susceptible to nucleophillic

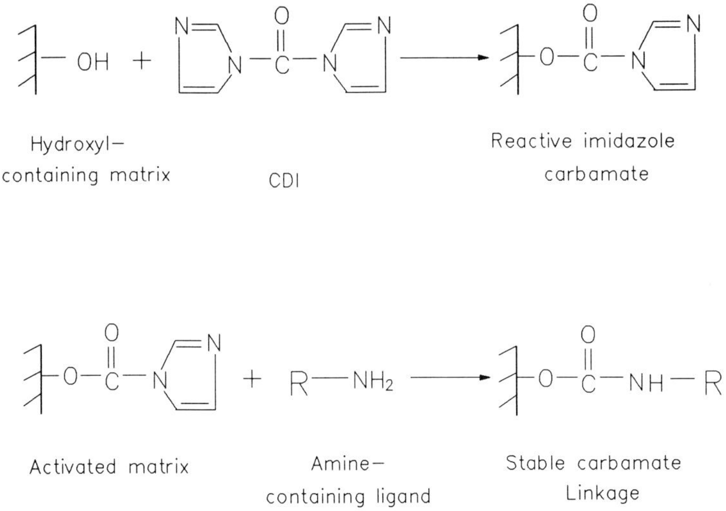

Figure 6. Activation of hydroxyl-containing solid phase with carbonyldiimidazole (CDI) and coupling of an amine-containing ligand.

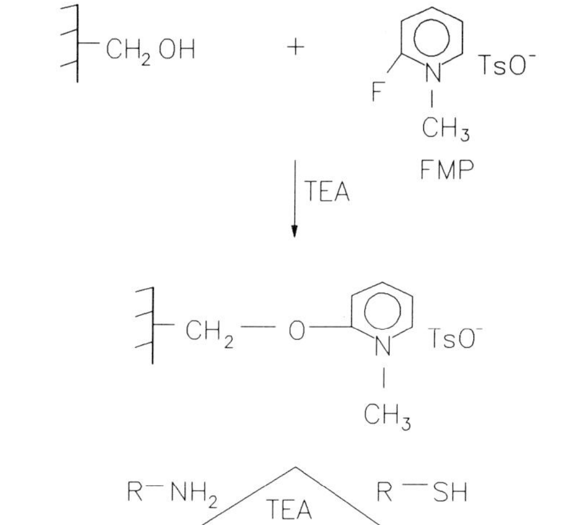

Figure 7. Activation of hydroxyl-containing solid phase with 2-fluoro-1-methylpyridinium toluene-4-sulfonate (FMP) and the coupling of an amine- or a thiol-containing ligand ($R\text{-}NH_2$ and R-SH, respectively). TsO^-, toluene-4-sulfonate; TEA, triethylamine.

attack by the ε-amino groups of lysine, result in a stable amide linkage under mild conditions with negligible leaching and no ionic properties (38) but instability of these adsorbents at alkaline pH has been reported (see Section 6). Epoxy-activated matrices have reactive epoxides at the end of long-chain hydrophilic spacer arms. This reactive group can be coupled to an amine, hydroxyl, or other nucleophilic ligand. Epoxy activation provides very stable bonding but generally requires a pH of 9–13 and temperatures of 20–45 °C, conditions which are incompatible with sensitive ligands (39): pH 9–11 and room temperature are generally used to couple amino- and thiol-containing ligands by this method.

CDI and FMP activation protocols are outlined below as the methods of choice. Also included is a protocol for immobilizing spacer arms to the support. The latter are linear hydrocarbons (e.g. 1,6-diaminohexane, 6-aminohexanoic acid, diaminodipropylamine) with functional amino and/or carboxyl groups on both ends. These spacers protrude from the matrix backbone and their terminal amino/carboxyl groups are used to couple the ligand. Depending upon the characteristics of a particular affinity binding site, use of spacers may be crucial for successful interaction between the ligand and the product (40).

Protocol 2. Carbonyldiimidazole (CDI) activation and coupling of the ligand (*Figure 6*)

Equipment and reagents

- Solid phase such as agarose or silica containing hydroxyl groups
- Non-aqueous polar solvent such as acetone or dioxane (HPLC-grade)
- CDI (Aldrich)
- Ligand dissolved in a suitable buffer (e.g. PBS). Use the ligand concentration required to give maximal adsorbent productivity[a]
- Imidazole (BDH), 2 mM in solution A
- Solution A: 0.15 M NaOH titrated to pH 7 with 0.15 M HCl
- Solution B: 0.06 M *p*-nitrobenzoyl chloride dissolved in acetone. *Prepare fresh*
- PBS containing 0.02% (w/v) sodium azide
- PBS (0.02 M potassium phosphate, pH 7.4, containing 0.15 M NaCl)
- 0.3 M ethanolamine (adjusted to pH 8.0 with HCl) or 0.2 M Tris–HCl, pH 9.0
- 0.15 M HCl
- 1 M NaCl
- 1 M $NaHCO_3$
- 1 M NaOH
- Apparatus for constant agitation (e.g. Multimix major; Luckham)
- Sintered glass funnel connected to a suction pump
- Spectrophotometer

A. *CDI activation*

Note: This procedure must be done in a fume cupboard if dioxane is used.

1. Wash the solid phase with several volumes of water to remove any preservatives.
2. Transfer the material into a non-aqueous phase such as acetone or dioxane.[b] To do this, wash the support on a sintered glass filter funnel

successively with solutions containing solvent and water in ratio 30:70 and 70:30 and finally with neat solvent.[c]

3. Resuspend in an equal volume of the solvent, add solid CDI to give a concentration of 1% (w/v), and activate the matrix by constant mixing for 1 h at room temperature.
4. Transfer the activated support to a filter funnel and wash with several volumes of solvent. The activated gel can be stored at −20 °C until required.[d]
5. Determine the activation level and couple the ligand to the activated support.

B. *Determination of activation level*

1. Suction-dry the support and hydrolyse 20–100 mg[e] in 2 ml of 0.15 M NaOH by agitating continuously at room temperature for 20 h. Centrifuge the mixture.
2. Neutralize the supernatnant with 0.15 M HCl and to aliquots (made up to 1 ml with solution A), add 0.1 ml of 1 M $NaHCO_3$ followed by 1 ml of solution B. Set up samples containing standard imidazole solution (0–2 μmol/ml) simultaneously.
3. Mix vigorously for a few seconds and leave at room temperature for 10 min.
4. Add 2 ml of 1 M NaOH, mix, and leave for 30 min at room temperature.
5. Read the absorbance at 415 nm and calculate the amount of imidazole (μmol/g drained matrix) in the test samples by reference to the standard curve.

C. *Ligand coupling*

1. For ligands that are soluble in and compatible with the solvent used for activation, suction-dried support can be used. To couple ligands which are sensitive to the solvent, transfer the support to water by washing sequentially with solutions containing 70% and 30% (w/v) of solvent and then with water.
2. Add the ligand dissolved in a suitable coupling buffer (e.g. PBS).[a,f] Allow the reaction to continue by constant agitation for 6 h at room temperature or 20 h at 4 °C.
3. Wash the coupled support with at least 5 volumes each of coupling buffer, 1 M NaCl, and water. Pool all the washings and determine the amount of ligand present. Calculate the density of the ligand coupled by the difference between the starting concentration and that recovered in the wash solutions. Absorbance at 280 nm can be used to obtain this value.

Protocol 2. *Continued*

4. Block the unreacted groups by suspending the gel in 1 volume of Tris–HCl (0.2 M, pH 9) or 0.3 M ethanolamine pH 8.0 and agitating constantly at room temperature for 2–4 h.
5. Wash the adsorbent extensively in 1 M NaCl and store the adsorbent as a slurry in buffer containing a preservative (e.g. 0.02% (w/v) NaN_3) at 4 °C.

[a]See Section 4.
[b]CDI is extremely unstable in aqueous solutions. In presence of water it dissociates immediately into CO_2 and imidazole. Therefore the activation step must be done in a solvent which is free of water.
[c]To ensure proper transfer, degas the solid phase (especially in the neat solvent) prior to the final wash.
[d]A hydroxyl-containing matrix that is CDI-activated is stable in non-aqueous solutions for a long time (more than a year in our experience).
[e]The weight required will depend upon the support used; for example, agarose-based matrices will have higher activation levels than silica-based matrices due to the porous nature of the latter.
[f]Avoid amine-containing buffers such as Tris and glycine as they will compete with the ligand. Neutral pH is recommended for the coupling reaction as the hydrolysis of the imidazole group is slower at this pH than at the pH 8.5–11 generally used.

Protocol 3. 2-fluoro-1-methyl pyridinium toluene-4-sulfonate (FMP) activation and ligand coupling[a] (*Figure 7*)

Warning: All the procedures should be carried out in a fume cupboard because of the volatile solvents and amines used in this protocol.

Equipment and reagents

- Solid phase, such as agarose or silica, containing hydroxyl groups
- Acetone
- Dry acetonitrile (Aldrich)
- FMP (Aldrich)
- Dry triethylamine (Aldrich)
- Solution A: dry acetonitrile containing 2% (v/v) dry triethylamine (*prepare fresh*)
- Ligand dissolved in 0.2 M sodium carbonate, pH 8.5
- 1 M NaCl
- 0.2 M sodium carbonate buffer, pH 8.5
- 2 mM HCl
- 0.2 M Tris–HCl, pH 9.0
- Apparatus for constant agitation (e.g. Multimix major; Luckham)
- Sintered glass funnel connected to a suction pump
- Spectrophotometer
- Oven at 100 °C

A. *FMP activation*

1. Wash the solid phase with several volumes of water to remove any preservatives.
2. Transfer the matrix to acetonitrile by successively washing with 25:75, 50:50, 75:25 acetone:water mixtures, and then with acetone and dry acetonitrile.
3. Place the gel in a beaker (dried at 100 °C) containing 1 volume of solution A and stir vigorously.

4. Dissolve FMP (0.6 g/10 ml adsorbent) in 2 volumes of solution A. Add the solution slowly to the gel suspension and stir for a further 10 min.
5. Transfer the activated gel to a sintered glass filter and wash successively with acetonitrile, acetone, 50:50 acetone:water mixture and 2 mM HCl. The activated gel is stable in 2 mM HCl for at least 4 months and can be stored at 4 °C.
6. Calculate the amount of FMP immobilized by determining the concentration of 1-methyl-2-pyridone in the pooled washes using the molar extinction at 297 nm of 5900.

B. *Coupling of ligand*

1. Wash the support with water and suction dry FMP-activated support.
2. Resuspend the adsorbent in 1 volume of 0.2 M sodium carbonate buffer, pH 8.5, containing the ligand to be coupled.
3. Allow the reaction to continue by constant agitation for 6 h at room temperature or 20 h at 4 °C.
4. Wash the coupled gel with 10 volumes each of sodium carbonate buffer, 1.0 M NaCl, and water to remove unreacted ligand. Determine the concentration of the ligand in the pooled wash either by absorption at 280 nm or using a protein assay. Calculate the density of the coupled ligand by the difference between the starting concentration and that recovered in the wash solutions.
5. Block the unreacted activated hydroxyl groups by suspending the gel in 1 volume of 0.2 M Tris–HCl, pH 9, and agitating constantly at room temperature for 2–4 h.
6. Wash the coupled support extensively with 1 M NaCl and store the adsorbent as a slurry at 4 °C in the salt solution containing a preservative (e.g. 20% (v/v) ethanol).

[a]Footnotes *a*, *c*, and *f* given in *Protocol 2* also apply here.

Protocol 4. Immobilization of diaminodipropylamine (DADPA) spacer molecules

Note. This procedure should be carried out in a fume cupboard and gloves should be worn when handling DADPA and ninhydrin.

Equipment and reagents

- CDI-activated solid phase transferred to acetone (see *Protocol 2*)
- DADPA (Sigma)
- Dry acetone (Aldrich)
- Ninhydrin reagent (Sigma)
- 1 M NaCl
- Ethanol
- 50% (v/v) aqueous ethanol

Protocol 4. *Continued*

- Apparatus for constant agitation (e.g. Multimix major; Luckham)
- Spectrophotometer
- Sintered glass funnel connected to a suction pump
- Water bath heated to boiling

A. *Coupling of DADPA*

1. Dissolve DADPA (20% w/v) in acetone.
2. Suction-dry the activated support and add 1 volume of DADPA solution.
3. Agitate the mixture continuously at room temperature for 4 h.
4. Wash the gel with at least 5 volumes of acetone, by suction, to remove excess DADPA. Transfer to water by washing with 50:50 acetone:water mixture followed by water, 1 M NaCl, and water. Check the final wash for the presence of primary amines to ensure complete removal of DADPA (see *Protocol 4B*).
5. Resuspend the gel in buffer for further coupling to a ligand[a] as described in *Protocol 2*.

B. *Ninhydrin assay for detecting primary amines*[b]

1. Suction dry the coupled gel and weigh suitable aliquots (20–50 mg) into test tubes. Resuspend in 2 ml of water. (To test the wash solutions, use aliquots made up to 2 ml with water.)
2. Add 1 ml of ninhydrin reagent and heat in a boiling water bath for 15 min.
3. Cool to ambient temperature and measure the absorbance at 570 nm. This value should be low in the wash solutions. For the activated matrix, prior dilution (30- to 50-fold) in 50% ethanol (which dissolves the purple colour) is normally required. Allow the support to settle and use the clear supernatant for measuring the absorbance.
4. Calculate the amount of amines present per gram of suction-dried gel using a molar extinction coefficient of 8750 for the amine–ninhydrin complex.

[a]Spacers containing diamines are suitable for coupling ligands which contain a carboxyl group. Use spacers such as 6-aminohexanoic acid if coupling through an amino group is required.
[b]This assay is suitable for both checking that the final wash is complete and determining the activation levels of the coupled gel.

6. Ligand leaching

High stability of matrix–ligand linkages is a prerequisite for the use of biospecific adsorbents in affinity chromatography. Despite the introduction of the various activation chemistries, ligand leaching remains a problem. Some of the studies documented in literature are described below. Non-covalent binding of ligand, support degradation, protein degradation, and bond solvolysis have all been implicated in the detachment of immobilized ligands though the actual mechanisms involved have been pinpointed in only a few cases. With CNBr-immobilized ligands, leakage is significant at room temperature if the pH is above 7 because the iso-urea bond is reversible and can be cleaved by hydrolysis at alkaline pH and by aminolysis with low molecular mass amines (41). Sundberg and Porath (39) showed that the linkage formed between epoxy-activated Sepharose 6B and glycine was stable at 70 °C and pH 11 for at least 8 weeks; at pH 7 about 10% of the bound amino acid was released after 5–6 weeks whereas at pH 3 the ligand leakage was continuous and more than 50% of the amino acid was lost after 8 weeks. In contrast, Koelsch *et al.* (42) demonstrated that epoxy derivatives of hydroxyalkyl methacrylate gels were more stable under these conditions: only 20% of the ligand was lost during long storage and this release was due to the splitting of the peptide bond adjacent to the covalent bond anchoring the peptide to the matrix. The authors attributed the low leakage to the presence of a sub-population of fixed ligands which were susceptible to matrix-induced bond splitting. Wilchek and Miron (43) have reported instability of *N*-hydroxy-succinimide-activated adsorbents at alkaline pH and their constant leakage during use. Van Sommeren *et al.* (44) showed that leakage from *N*-hydroxy-succinimide-activated support was considerably less than that with tresyl chloride- or hydrazide-activated adsorbents. Leckband and Langer (45) showed that leakage of lysozyme, bovine serum albumin, and heparinase from *N*-hydroxysuccinimide- or trifluoroethanesulfonic acid-activated agarose support (prepared commercially) was predominantly due to the loss of non-covalently bound protein. These effects could be minimized by washing the adsorbent thoroughly, after the ligand coupling reaction, using buffers containing a high concentration of salt (1 M NaCl) prior to blocking of the un-reacted sites. Ligand leakage from eight different commercially available Protein-A matrices has also been reported (46). Conversely, a novel affinity matrix prepared by covalently binding Protein A to crystalline cell surface layers from the Gram-positive *Clostridium thermohydrosulphuricum* has been described recently as having extremely low leakage (47).

7. Ligand orientation and performance of affinity adsorbents

Most of the activating chemistries listed above couple to the ligand randomly through amino residues. Such random coupling through amino acids is often accompanied with a partial or complete loss of binding capacity owing to the reduced efficiency of ligand–product binding. One reason for this is the

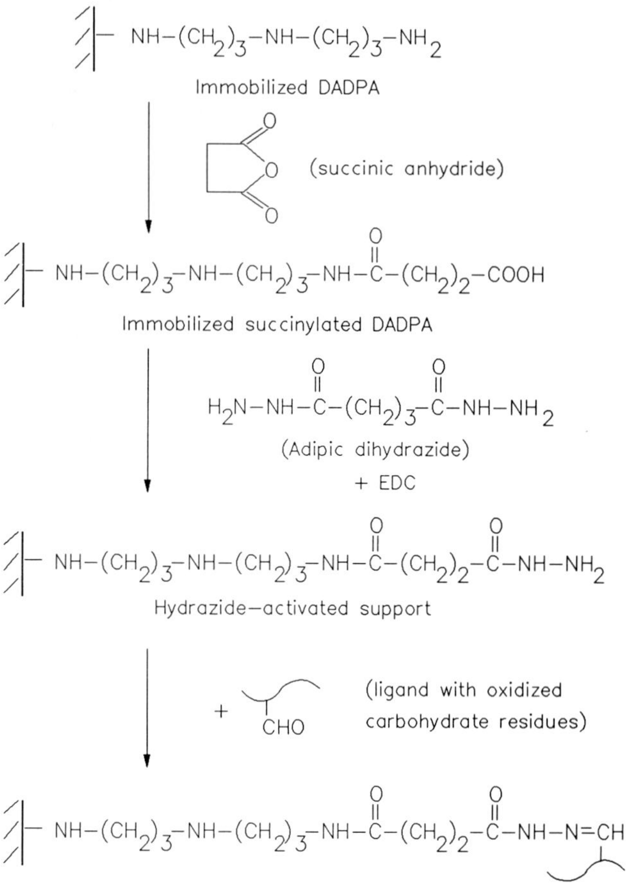

$NH{-}(CH_2)_3{-}NH{-}(CH_2)_3{-}NH_2$

Immobilized DADPA

(succinic anhydride)

$NH{-}(CH_2)_3{-}NH{-}(CH_2)_3{-}NH{-}C(=O){-}(CH_2)_2{-}COOH$

Immobilized succinylated DADPA

$H_2N{-}NH{-}C(=O){-}(CH_2)_3{-}C(=O){-}NH{-}NH_2$

(Adipic dihydrazide)

+ EDC

$NH{-}(CH_2)_3{-}NH{-}(CH_2)_3{-}NH{-}C(=O){-}(CH_2)_2{-}C(=O){-}NH{-}NH_2$

Hydrazide–activated support

+ CHO (ligand with oxidized carbohydrate residues)

$NH{-}(CH_2)_3{-}NH{-}(CH_2)_3{-}NH{-}C(=O){-}(CH_2)_2{-}C(=O){-}NH{-}N{=}CH$

Figure 8. Preparation of hydrazide gel and coupling of ligand containing carbohydrate residues.

multisite attachment of the ligand, if the activated support contains an excess of activation groups, making the binding site inaccessible to the product due to steric constraints. More recently, the random orientation of the coupled ligand has been documented as a significant factor contributing to the optimum performance of an affinity adsorbent (48). This is particularly the case when ligands such as monoclonal antibodies or catalytic antibodies ('abzymes') are used. In these cases, maintaining conformational integrity of the binding site upon immobilization is crucial to produce an active immunosorbent. Various methods have recently been developed to facilitate the correct orientation. These include:

- oxidation and subsequent immobilization through the carbohydrate region
- immobilization via the carboxyl group

- immobilization via sulfhydryl groups (thiophilic adsorption chromatography) (see Chapter 3)
- metal ion interaction between the side chains of amino acids exposed on the surface of the proteins (immobilized metal ion affinity chromatography) (see Chapters 3 and 7)
- use of designer mimetic ligands (see Chapter 3)
- engineering of fusion proteins and protein fragments to insert novel ligands for use in affinity chromatography (see Chapter 7)

7.1 Oxidation and subsequent immobilization through the carbohydrate region

The immobilization of glycoproteins through their carbohydrate moieties is useful for enzymes and antibodies because their active complementary binding sites are located on their protein moieties. In the case of immunoglobulins the carbohydrates are located mainly on the Fc region of the molecules; binding through this region therefore ensures the exposure of the Fab domains for antigen binding. Advantages of immobilizing enzymes through their carbohydrate components have been reviewed (49). In these cases the carbohydrates are first oxidized by periodate and then coupled to an immunoadsorbent activated by hydrazide (50). Hydrazide-activated supports are available commercially, e.g. Affigel (Bio-Rad), polyacrylhydrazido-agarose (Sigma), and agarose adipic acid hydrazide (Pharmacia). Preparation of hydrazide gels and coupling of ligands is described below (see *Figure 8*).

Protocol 5. Preparation of hydrazide gel and coupling of ligand (*Figure 8*)

Equipment and reagents

- CDI-activated gel coupled to DADPA (see *Protocol 4*) or commercial aminated support
- EDC [1-ethyl-3-(3-dimethylaminopropyl) carbodiimide] (Aldrich)
- Succinic anhydride (Aldrich)
- 5 M NaOH
- 0.1 M Mes buffer, pH 4.75
- 0.5 M adipic dihydrazide in Mes buffer (Aldrich; prepare fresh in fume cupboard)
- Sodium *meta*-periodate (Aldrich)
- 0.1 M sodium phosphate buffer, pH 7.0
- Ligand for coupling (containing carbohydrate residues)
- 1 M NaCl
- Apparatus for constant agitation (e.g. Multimix major; Luckham)
- Sintered glass funnel connected to a suction pump
- Sephadex G-25 (Pharmacia); 10 ml packed in 1 cm × 10 cm column
- Detector (e.g. Uvicord, Pharmacia) to monitor eluate from G-25 column at 280 nm

A. *Preparation of succinylated adsorbent*

1. Wash the aminated support with at least 5 volumes of water to remove any preservatives and resuspend in 1 volume of water.
2. Add succinic anhydride (1 g/10 ml gel) and agitate constantly for 1 h at room temperature. Maintain pH at 6.0 with 0.5 M NaOH.

Protocol 5. *Continued*

3. Wash the gel extensively by suction with water, 1 M NaCl, and water.
4. Test a sample for the presence of primary amines (see *Protocol 4*). A negative test indicates complete succinylation. If a positive colour is obtained repeat steps 2 and 3.

B. *Preparation of hydrazide-activated support*

Note: use a fume cupboard for this procedure.

1. Wash the succinylated gel with 5 volumes of Mes buffer and suction dry.
2. Add 1 gel volume of adipic dihydrazide solution followed by 0.5 g of EDC. Agitate constantly for 3 h at room temperature.
3. Wash the gel with 5 volumes each of water, 1 M NaCl, water, and phosphate buffer. Suction dry and use to couple oxidized ligand as described below.

C. *Oxidation of glycoproteins*

1. Dissolve the glycoprotein (1–10 mg) in 1 ml of phosphate buffer.
2. Add 10 mg sodium *meta*-periodate and mix. Incubate at room temperature **in the dark** for 30 min.
3. Desalt the solution on the G-25 column previously equilibrated in phosphate buffer and collect the protein peak.

D. *Coupling of the oxidized ligand to hydrazide gel*

1. Mix the G-25[a] eluate, diluted appropriately in phosphate buffer, with an equal volume of hydrazide gel and agitate for 4 h at room temperature.
2. Wash the gel extensively with phosphate buffer, 1 M NaCl, then again with phosphate buffer. Store at 4°C resuspended in phosphate buffer containing 0.02% sodium azide.

[a] Use the ligand concentration required to give maximal adsorbent productivity. See Section 4C.

7.2 Immobilization via carboxyl group

Immobilization of proteins via their carboxyl group can be achieved by a water-soluble carbodiimide-mediated coupling reaction between an amine-containing support and a carboxyl group of the ligand. Aminated agaroses are available from Sigma, Bio-Rad, and Pierce. Other supports can be prepared easily by activating them with, e.g., CDI followed by the coupling of spacers containing diamino groups (diaminohexane, diaminodipropylamine) as described in *Protocol 6*.

Protocol 6. Immobilization of ligand through carboxyl groups

Equipment and reagents

- CDI-activated gel coupled to DADPA (see *Protocol 4*) or commercial aminated support
- EDC [1-ethyl-3-(3-dimethylaminopropyl) carbodiimide] (Aldrich)
- *N*-hydroxysulfosuccinimide (sulfo-NHS; Aldrich)
- 0.1 M sodium phosphate buffer, pH 7.0
- Ligand dissolved in phosphate buffer[a]
- 1 M NaCl
- 0.2 M sodium acetate
- Acetic anhydride (Aldrich)
- 0.1 M NaOH
- Apparatus for constant agitation (e.g. Multimix major; Luckham)
- Sintered glass funnel connected to a suc-

Method

1. Wash the aminated support with at least 5 volumes of water to remove any preservatives followed by 5 volumes of phosphate buffer.
2. Resuspend the gel in 1 volume of buffer containing dissolved ligand.
3. Add EDC (3% w/v) and sulfo-NHS[b] (0.5% w/v), mix, and agitate constantly at room temperature for 4 h.
4. Wash the coupled gel successively with phosphate buffer, water, 1 M NaCl, and water. Calculate the density of the coupled ligand by the difference between the starting concentration and that recovered in the wash solutions. Absorbance at 280 nm can be used to obtain this value.
5. Block the unreacted sites: suction dry the gel, resuspend in 1 volume of 0.2 M sodium acetate, and add one-half the gel volume of acetic anhydride.[c] Agitate constantly at 4°C for 30 min.
6. Add a further one-half the gel volume of acetic anhydride and mix the suspension at room temperature for 30 min.
7. Wash the blocked gel on a sintered glass funnel with water, 0.1 M NaOH, and water in succession. Store at 4°C resuspended in phosphate buffer containing 0.02% sodium azide.

[a]Use the ligand concentration required to give maximal adsorbent productivity. See Section 4.
[b]If the ligand contains both amino and carboxyl groups omit sulfo-NHS; otherwise, the protein may precipitate.
[c]Handle acetic anhydride in a fume cupboard.

7.3 Immobilization via sulfhydryl group

This method is useful for immobilizing ligands containing sulfhydryl groups which do not interfere with the interaction between the product and the ligand (51,52). Two types of method can be used to achieve purification of proteins of interest:

(a) Thio-gels are prepared by coupling sulfhydryl compounds such as β-mercaptoethanol or dithiothreitol (DTT) to a support. Such supports

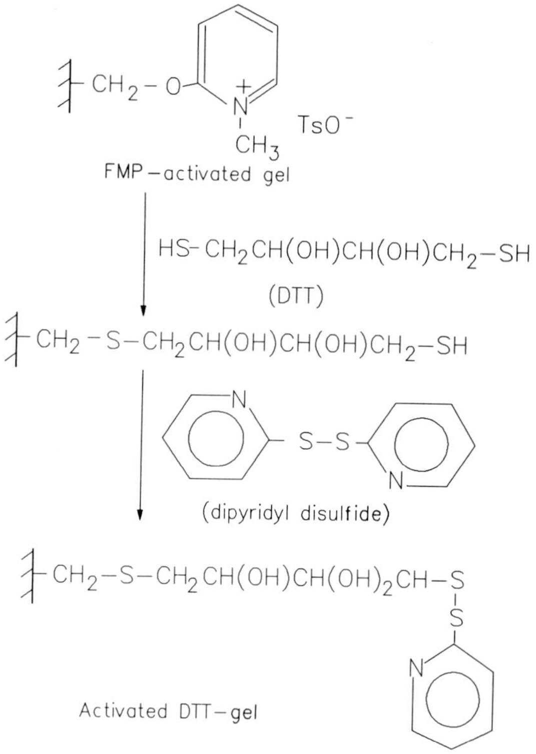

Figure 9. Preparation and activation of a thio-gel.

(available commercially from, e.g., Pharmacia) can be used for the purification of sulfhydryl-containing proteins. The purification proceeds through two distinct steps. First, the -SH group of the protein undergoes a thiol–disulfide interchange reaction. As a result of the interchange reaction, 2-thiopyridone is released and simultaneously the protein is covalently bound to the gel via a disulfide bond. Secondly, the bound protein is released by reduction of disulfide bonds with a reducing agent such as DTT. Thio-gels hence react spontaneously and reversibly under mild conditions and are simple to reactivate and reuse. Thio-gels can be prepared in-house by, e.g., reacting FMP-activated gel with DTT (see *Protocol 7*; *Figure 9*)

(b) Ligands containing sulfhydryl groups can be immobilized to obtain permanent coupling and used to purify the corresponding specific components. This technique has found a unique application in the purification of immunoglobulins because, contrary to other affinity purification methods, it adsorbs three major classes of immunoglobulins and their subclasses (53,54). FMP-activated gels (*Protocol 3*) and iodoacetyl-activated gels can be used to couple the ligands (see *Protocol 8*; *Figure 10*).

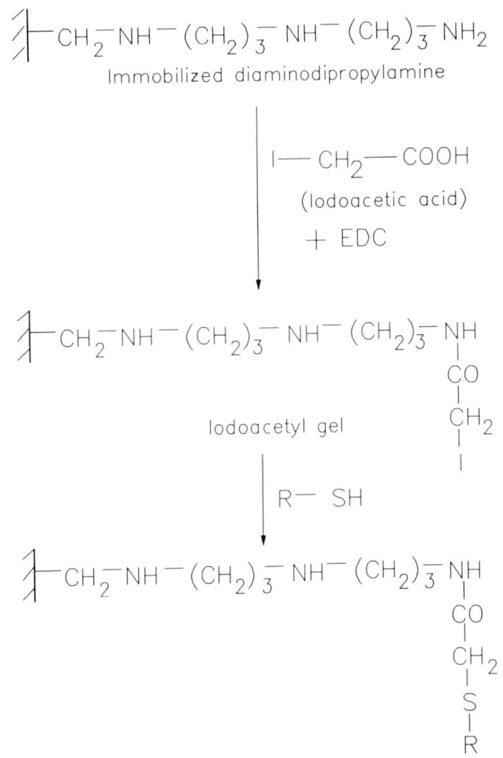

$-CH_2-NH-(CH_2)_3-NH-(CH_2)_3-NH_2$

Immobilized diaminodipropylamine

$I-CH_2-COOH$

(Iodoacetic acid)

+ EDC

$-CH_2-NH-(CH_2)_3-NH-(CH_2)_3-NH-CO-CH_2-I$

Iodoacetyl gel

R–SH

$-CH_2-NH-(CH_2)_3-NH-(CH_2)_3-NH-CO-CH_2-S-R$

Figure 10. Preparation of a solid phase with immobilized iodoacetyl group. EDC, 1-ethyl-3-(3-dimethylaminopropyl)carbodiimide; R-SH, protein containing sulfhydryl group.

Protocol 7. Preparation, activation, and use of thio-gel (*Figure 9*)

Equipment and reagents

- FMP-activated gel (see *Protocol 3*)
- Dithiothreitol (DTT: Sigma): prepare fresh 1 M solution in 0.2 M $NaHCO_3$
- 2,2′-dipyridyl disulfide (Aldrich)
- 1 mM EDTA
- 0.2 M $NaHCO_3$
- 2 mM HCl
- Assay buffer: 0.1 M sodium phosphate pH 8, containing 1 mM EDTA
- Ellman's reagent, 5,5′-dithiobis (2-nitrobenzoic acid) (Sigma) dissolved in assay buffer (4 mg/ml)
- Buffer 1: 0.05 M Tris–HCl, pH 7.4, containing 0.5 mM EDTA
- Buffer 2: 0.05 M Tris–HCl, pH 7.4, containing 0.05 mM EDTA
- Buffer 3: buffer 2 containing 1 M NaCl
- Buffer 4: buffer 2 containing 20 mM DTT
- Acetone
- Apparatus for constant agitation (e.g. Multimix major; Luckham)
- Sintered glass funnel connected to a suction pump
- Spectrophotometer
- Microcentrifuge
- Sephadex G-25; 10 ml packed in a 1 cm × 10 cm column to desalt a 3 ml sample
- Detector (e.g. Uvicord, Pharmacia) to monitor column eluates at 280 nm

Protocol 7. *Continued*

A. *Preparation of thio-gel*

1. Wash FMP-activated gel with water on the glass filter to remove any preservatives and suction dry.
2. Resuspend in 1 gel volume of 0.2 M $NaHCO_3$ containing 1 M DTT and agitate constantly at room temperature for 5 h.
3. Wash the gel successively with at least 5 volumes of 0.2 M $NaHCO_3$s water, 2 mM HCl, and water. Determine the thiol content as described below.

B. *Determination of activation levels*

1. Wash the thio-gel with 5–10 volumes of assay buffer on the glass filter and suction dry.
2. Weigh aliquots (20–50 mg) in microfuge tubes and add 1 ml of assay buffer followed by 100 μl of Ellman's reagent. Mix at room temperature for 15 min.
3. Centrifuge the suspension (10 000 *g*, 5 min) and measure the absorbance of the supernatant at 412 nm. Calculate the thiol content per gram of suction-dried gel using the molar extinction coefficient of 13 600.

C. *Activation of thio-gel*

1. Transfer the thio-gel to 50% (v/v) acetone in water by washing on the glass funnel.
2. For each ml of gel add 100 mg of 2,2′-dipyridyl disulfide dissolved in 1 ml of 50% acetone. Stir constantly at room temperature for 30 min.
3. Wash the activated support with 50% acetone to remove excess 2,2′-dipyridyl disulfide (yellow until the wash, due to 2-thiopyridone released during the reaction, becomes colourless).
4. Wash and store the gel in 1 mM EDTA.

D. *Purification of sulfhydryl-containing proteins*

1. Pack 5 ml of the activated gel in a column and equilibrate the column in an appropriate buffer (e.g. 0.05 M Tris–HCl, pH 7.4) containing 0.5 mM EDTA (buffer 1).
2. Equilibrate the protein sample in buffer 1 by either dialysis or desalting on a Sephadex G-25 column previously equilibrated in this buffer.[a]
3. Apply the protein solution to the column and wash with buffer 1 until the absorbance of the eluate at 280 nm is less than 0.1.

4. Wash the column further with buffer 3 to remove any non-specifically adsorbed material until an absorbance at 280 nm of less than 0.02 is obtained.
5. Elute the protein by applying buffer 4.
6. Wash the column successively with 0.2 M $NaHCO_3$, water, 2 mM HCl, and water. Reactivate the gel for reuse (see *Protocol 7c*).

[a]The protein sample should be free of sulhydryl compounds as these would compete with the protein for binding sites on the adsorbent.

Protocol 8. Preparation of iodoacetyl-activated support (*Figure 10*)

Equipment and reagents

- CDI-activated support coupled to DADPA (see *Protocol 4*) or commercial aminated support
- Ligand containing sulfhydryl groups dissolved in PBS[a]
- PBS (0.02 M potassium phosphate, pH 7.4, containing 0.15 M NaCl)
- Iodoacetic acid (Sigma): prepare fresh a 15% (w/v) solution in water and adjust the pH to 4.5 with 1 M NaOH
- EDC (Aldrich)
- 1 M NaOH
- 1 M HCl
- 10 mm EDTA
- 1 M NaCl
- Apparatus for constant agitation (e.g. Multimix major; Luckham)
- Sintered glass funnel connected to a suction pump

Method

1. Wash the DADPA support with water to remove any preservatives and suction dry.
2. Suspend in 1 volume of iodoacetic acid solution and mix.
3. Add EDC (10% w/v) to the mixture and agitate constantly at room temperature for 4 h, adjusting the pH to 4.5 with HCl or NaOH, particularly in the first hour of reaction.
4. Wash the activated gel extensively with water followed by 1 M NaCl and water. Store in 10 mM EDTA at 4°C protected from light.
5. Couple the sulfhydryl containing ligand, dissolved in PBS, by reacting with the activated support for 4–6 h at 4°C. Wash the support with coupling buffer, 1 M NaCl, and water and store as a suspension in PBS containing 0.02% (w/v) sodium azide.

[a]Use the ligand concentration required to give maximal adsorbent productivity. See Section 4c.

7.4 Immobilized metal-ion affinity chromatography (IMAC)

In 1975 Porath *et al.* (55) reported a new method for the purification of proteins, making use of the selective retention of proteins on transition metal ions chelated to an insoluble matrix such as agarose. The interaction occurs between the side-chains of particular amino acids (histidine, cysteine, and tryptophan) exposed on the surface of protein and divalent metal ions of copper, cobalt, nickel, and zinc. This technique has been used to purify various proteins. In this technique the gel is first charged with the metal ions to form the chelate. Proteins and other biomolecules will bind to the gel depending upon the presence of the surface groups which have affinity for chelated metal. The binding strength is mainly affected by the nature of the metal ion and the pH of the buffers. Since the metal ions are strongly bound to the matrix the adsorbed proteins are eluted either by competitive elution, lowering the pH, or the use of strong chelating agents.

Hence the important requirement of this technique is that the metal ion must be orientated so as to maximize its interaction with the protein. To achieve this at least two modifications have been introduced to the standard method:

(a) Synthetic peptides are designed as biospecific ligands for coupling to the solid phase. The metal binding domain of the protein is identified from the known amino acid sequence of the protein. The peptide is synthesized and immobilized on the support using chemical coupling procedures consistent with the retention of metal binding properties. The support is then charged with the metal and used to bind the protein of interest (56,57).

(b) A novel procedure for orientated immobilization of monoclonal antibodies was described by Loetscher *et al.* (58). It combines site-specific covalent modification of the carbohydrate moiety of immunoglobulins with Ni^{2+}-chelate affinity chromatography. A hexahistidine peptide (chelating peptide) was chemically conjugated to the aldehyde groups generated on the carbohydrate side-chains of the monoclonal antibodies. The selective interaction of histidine-containing peptide with Ni^{2+} resulted in an orientated immobilization of monoclonal antibody which retained most of the binding activity.

Commercial supports with immobilized iminodiacetate, available from various sources (e.g. Pharmacia, Pierce, and Sigma), are suitable for use in metal ion affinity chromatography. See Chapters 3 and 7 for further details of IMAC.

7.5 Use of Cibacron Blue F3G-A and designer mimetic ligands

The commercially available textile dye Cibacron Blue F3G-A can be readily coupled to a variety of support matrices and has been used for the purification of many enzymes that require adenylic cofactors such as NAD^+, $NADP^+$, and ATP (59). It has been shown to be involved in the binding in the NAD^+-binding site of horse liver alcohol dehydrogenase through the anthraquinone and the diaminobenzene rings (*Figure 11*; cited in ref. 60). Coupling of a

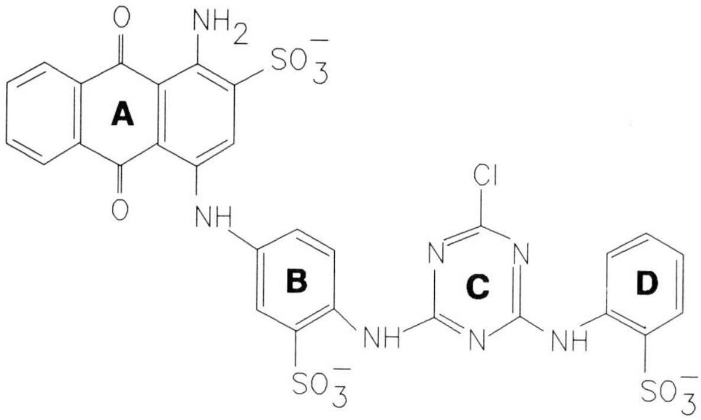

Figure 11. The structure of Cibacron Blue F3G-A. Anthraquinone (A) and diaminobenzene (B) rings are involved in the binding of enzymes which require adenylic cofactors such as NAD^+. Coupling of a support through the triazine ring (C) provides a ligand with correct orientation for coupling to the enzymes. D is the terminal ring of the molecule.

hydroxyl-containing support through the triazine ring would therefore provide a ligand in the correct orientation for efficient binding of the enzymes to be purified.

Lowe *et al.* (60) have used computer-aided molecular design to develop a number of designer dyes which mimic the structure and binding of natural biological ligands. The resulting dyes produce novel, highly selective, biomimetic ligands which display a higher affinity for the target enzymes than the parent dyes. Ten mimetic ligand adsorbents are marketed by Affinity Chromatography. Limited A 'Protein Isolation Kit for Sorbent Identification' containing the ten adsorbents is available for small-scale purifications to select the adsorbent most suitable for the protein of interest. See Chapter 3 for further details of dye affinity separations.

7.6 Engineered fusion proteins

Several systems which facilitate the correct orientation of ligand/product, including Protein A and metal chelating peptides, have been devised. Protein A, a *Staphylococcus aureus* cell wall protein, shows a specific affinity for the Fc region of immunoglobulin G (IgG). This property has been exploited for the purification of recombinant proteins. Hybrid proteins have been prepared by fusing the coding sequence of the protein of interest with the coding sequence of staphylococcal Protein A or synthetic polypeptides based on domains B and A-B (61,62) together with a specific cleavage sequence. Such fusion proteins can be purified by taking advantage of the specific binding of Protein A affinity tail to immobilized IgG. After purification the affinity tail is cleaved off at the designed cleavage site and removed by techniques such as gel filtration to obtain the purified protein.

For metal chelate affinity chromatography, hybrid proteins comprising the protein of interest and polyhistidine affinity peptide fused at the C- or N-terminus have been successfully immobilized on adsorbents (25,63,64). See Chapter 7 for a more complete discussion of these methods.

8. Concluding remarks

An attempt has been made in this chapter to describe recent developments in affinity chromatography. Though only a limited number of protocols are given here, the approach adopted has been to address the parameters which affect the efficacy of the technique rather than just list the enormous number of methods which have become available in recent years.

Mention should be made of other affinity-based techniques which have emerged in the past few years. These offer alternatives to affinity chromatography and are particularly attractive, because of their scalability, for use in industrial processes. These include affinity precipitation, affinity aqueous two-phase extraction (65,66, respectively; mentioned briefly in Section 3.2), affinity ultrafiltration (67; also see Section 3.1.4), affinity electrophoresis (68), and affinity-based reverse micellar extraction (69).

References

1. Porath, J., Axen, R., and Ernback, S. (1967). *Nature*, **215**, 1491.
2. Lowe, C. R. and Dean, P. D. G. (1974). *Affinity chromatography*. John Wiley and Sons, London.
3. Dean, P. D. G., Johnson, W. S., and Middle, F. A. (eds) (1985). *Affinity chromatography: a practical approach.* IRL Press, Oxford.
4. Hill, C. R. (1990). In: *Separations for biotechnology 2* (ed. L. Pyle), pp. 431–2. Elsevier Applied Science, London.
5. Wun, T.-C., Schleuning, W.-D., and Reich, E. (1982). *J. Biol. Chem.*, **257**, 3276.
6. Horstmann, B. J. and Chase, H. A. (1989). *Chem. Eng. Res. Des.*, **67**, 243.
7. Mohan, S. B. and Lyddiatt, A. (1992). *Biotechnol. Bioengng*, **40**, 549.
8. Buckle, P. E., Davies, R. J., Kinning, T., Yeung, D., Edwards, P. R., Pollard-Knight, D., and Lowe, C. R. (1993). *Biosensors Bioelectronics*, **8**, 355.
9. Giddings, J. C. (1991). *Unified separation science.* John Wiley and Sons, New York.
10. Ohlson, S., Hansson, L., Larsson, P.-O., and Mosbach, K. (1978). *FEBS Lett.*, **93**, 5.
11. Afeyan, N. B., Fulton, S. P., and Regnier, F. E. (1991). *LC-GC Intl*, **4**, 14.
12. Stewart, D. J., Purvis, D. R., and Lowe, C. R. (1990). *J. Chromatogr.*, **510**,177.
13. McCreath, G. E., Chase, H. A., Purvis, D. R., and Lowe, C. R. (1992). *J. Chromatogr.*, **597**, 189.
14. Rigney, M. P., Funkenbusch, E. F., and Carr, P. W. (1990). *J. Chromatogr.*, **499**, 291.
15. Wirth, H.-J. and Hearn, M. T. W. (1993). *J. Chromatogr.*, **646**, 143.
16. Brandt, S., Goffe, R. A., Kessler, S. B., O'Connor, J. L., and Zale, S. E. (1988). *Bio/Technology*, **6**, 779.

17. Champluvier, B. and Kula, M.-R. (1992). *Biotechnol. Bioengng*, **40**, 33.
18. Iwata, H., Saito, K., and Furasaki, S. (1991). *Biotechnol. Prog.*, **7**, 412.
19. Serafica, G. C., Pimbley, J., and Belfort, G. (1994). *Biotechnol. Bioengng*, **43**, 21.
20. O'Brien, S. M., Thomas, O. R. T., and Dunnill, P. (1994). *Proceedings of the Institute of Chemical Engineers Research Event*, Vol. **1**, p. 159. Institute of Chemical Engineers, Rugby, UK.
21. Sebba, F. (1987). *Foams and biliquid foams: aphrons.* John Wiley and Sons, London.
22. Morris, J. E., Hoffman, A. S., and Fisher, R. R. (1993). *Biotechnol. Bioengng*, **41**, 991.
23. Johansson, G. (1986). In: *Partitioning in aqueous two-phase systems* (ed. H. Walter., D. E. Brooks, and D. Fisher), p. 161–226. Academic Press, New York.
24. Brunner, K. L. and Hemfort, H. (1988). In: *Advances in biotechnological processes*, Vol. 8 (ed. A. Mirzrahi), p. l. Alan Liss, New York.
25. Hansson, M., Stahl, S., Hjorth, R., Uhlen, M., and Moks, T. (1994). *Bio/Technology*, **12**, 285.
26. Wells, C. M., Lyddiatt, A., and Patel, K. (1987). In: *Separations for biotechnology 1* (ed. M. S. Verrall and M. J. Hudson), p. 217–224. Ellis-Harwood, Chichester, UK.
27. van Brakel, J. and Kleizen, H. H. (1990). In: *Critical reports in applied chemistry*, Vol. 29 (ed. M. A. Winkler), p. 95. Elsevier Applied Science, London.
28. Morton, P. and Lyddiatt, A. (1992). In: *Ion exchange advances* (ed. M. J. Slater), p. 237–244. Elsevier Applied Science, London.
29. Gilchrist, G. R., Burns, M. T., and Lyddiatt, A. (1994). In: *Separations for biotechnology 3* (ed. L. Pyle), p. 186–192. Royal Society of Chemistry, Cambridge, UK.
30. Gibson, N. B. and Lyddiatt, A. (1993). In: *Cellulosics: materials for selective separations and other technologies* (ed. J. F. Kennedy, G. O. Phillips, and P. A. Williams), p. 52–62. Ellis-Harwood, Chichester, UK.
31. Ritchi, H. J., Ross, P., and Woodword, D. R. (1990). *Chromat. Anal.*, **October 1990**, 9.
32. Chappell, I. (1988). *Lab. Pract.*, **38**, 59.
33. Mohan, S. B. and Lyddiatt, A. (1992). *J. Chromatogr.*, **584**, 23.
34. March, S. C., Parikh, I., and Cuatrecasas, P. (1974). *Anal. Biochem.*, **60**, 149.
35. Bethell, G. S., Ayers, J. S., Hancock, W. S., and Hearn, M. T. W. (1979). *J. Biol. Chem.*, **254**, 2572.
36. Nilsson, K. and Mosbach, K. (1980). *Eur. J. Biochem.*, **112**, 397.
37. Ngo, T. T. (1986). *Bio/Technology*, **4**, 134.
38. Wilchek, M. and Miron, T. (1985). *Appl. Biochem. Biotechnol.*, **11**, 191.
39. Sundberg, L. and Porath, J. (1974). *J. Chromatogr.*, **90**, 87.
40. Cuatrecasas, P. (1970). *J. Biol. Chem.*, **245**, 3059.
41. Wilchek, M., Oka, T., and Topper, Y. (1975). *Proc. Natl Acad. Sci. USA*, **72**, 1055.
42. Koelsch, R., Lasch, J., Blaha, K., and Turkova, J. (1984). *Enz. Microb. Technol.*, **6**, 31.
43. Wilchek, M. and Miron, T. (1987). *Biochemistry*, **26**, 2155.
44. Van Sommeren, A. P. G., Machielsen, P. A. G. M., and Gribnau, T. C. J. (1993). *J. Chromatogr.*, **639**, 23.
45. Leckband, D. and Langer, R. (1991). *Biotechnol. Bioengng*, **37**, 227.

46. Fuglistaller, P. (1989). *J. Immunol. Methods*, **124**, 171.
47. Weiner, C., Sara, M., and Sleytr, U. B. (1994). *Biotechnol. Bioengng*, **43**, 321.
48. Spitznagel, T. M. and Clark, D. S. (1993). *BioTechnology*, **11**, 825.
49. Turkova, J., Vohnik, S., and Helusova, S. (1992). *J. Chromatogr.*, **597**, 19.
50. O'Shannessy, D. J. and Wilchek, M. (1990). *Anal. Biochem.*, **191**, 1.
51. Porath, J., Maisano, F., and Belew, M. (1985). *FEBS Lett.*, **185**, 306.
52. Porath, J. and Hutchens, T. W. (1987). *Int. J. Quant. Chem.: Quant. Biol. Symp.*, **14**, 297.
53. Hutchens, T. W., Magnuson, J. S., and Yip, T.-T. (1990). *J. Immunol. Methods*, **128**, 89.
54. Nopper, B., Kohen, F., and Wilchek, M. (1989). *Anal. Biochem.*, **180**, 66.
55. Porath, J., Carlsson, J., Olsson, I., and Belfrage, G. (1975). *Nature*, **258**, 598.
56. Hutchens, T. W., Nelson, R. W., Li, C. M., and Yip, T.-T. (1992). *J. Chromatogr.*, **604**, 125.
57. Hutchens, T. W. and Yip, T.-T. (1992). *J. Chromatogr.*, **604**, 133.
58. Loetscher, P., Mottlau, L., and Hochuli, E. (1992). *J. Chromatogr.*, **595**, 113.
59. Clonis, Y. D., Atkinson, A., Breton, C. J., and Lowe, C. R. (1987). *Reactive dyes in protein and enzyme technology*. Stockton Press, New York.
60. Lowe, C. R., Burton, S. J., Burton, N. P., Alderton, W. K., Pitts, J. M., and Thomas, J. A. (1992). *Trends Biotechnol.*, **10**, 442.
61. Nilsson, B., Moks, T., Jansson, B., Abrahmsen, L., Elmblad, A., Holmgren, E., Henrichson, C., Jones, T. A., and Uhlen, M. (1987). *Protein Engng*, **1**, 107.
62. Kihira, Y. and Aiba, S. (1992). *J. Chromatogr.*, **597**, 277.
63. Hochuli, E., Bannwarth, W., Dobeli, H., Gentz, R., and Stuber, D. (1988). *Bio/Technology*, **6**, 1321.
64. Stuber, D., Matile, H., and Garotta, G. (1990). *Immunol. Methods*, **4**, 121.
65. Mattiasson, B., Senstad, C., and Ling, T. G. (1989). In: *Bioprocess engineering* (ed. T. K. Ghosh), p. 235–245. Ellis Harwood, Chichester, UK.
66. Franco, T., Andrews, B. A., Asenjo, J. A., Cascone, O., Hodgsdon, C., and Andrews, A. T. (1990). In: *Separations for biotechnology 2* (ed. L. Pyle) p. 335–344. Elsevier Applied Science, London.
67. Mattiasson, B. and Ramstrop, M. (1983). In: *Biochemical Engineering III*, (ed. K. Venkatasubramanian and A. Constantinid), p. 307–9. New York Academy of Sciences, New York.
68. Shimura, K. (1990). *J. Chromatogr.*, **510**, 251.
69. Woll, J. M., Hatton, T. A., and Yarmush, M. L. (1989). *Biotechnol. Prog.*, **5**, 57.

2

Quantitative affinity chromatography

DONALD J. WINZOR

1. Introduction

Quantitative affinity chromatography was introduced (1) at a stage when the preparative technique was well established as a method of solute purification, and has been developed to take additional advantage of the chromatographic matrix used for isolation of the solute on the basis of biospecificity. Whereas the function of the immobilized ligand in preparative affinity chromatography is the selective interaction with a particular solute, its role in quantitative affinity chromatography is to provide competition for the ligand whose biospecificity it is mimicking, and thereby to provide a means of determining the binding constant for the biospecific interaction. The topic under review is far broader than the characterization of interactions by column chromatography, which is merely one means of evaluating the perturbing effect of ligand on the partition coefficient describing the distribution of solute between soluble and adsorbed states.

The technique was certainly developed initially as a column procedure (1,2) whereby ligand-effected variation in the partition of solute between soluble and immobilized forms was monitored by the consequent change in elution volume of the solute. However, it has also been adapted so that the position of the partition equilibrium is ascertained directly by measuring the solute concentration in the liquid phase of a biphasic mixture with known total concentration (3). Such partition equilibrium studies have been rendered more accurate by the introduction of a recycling partition technique (4) in which all mixtures are prepared by successive additions of solute to one sample (slurry) of affinity matrix. Direct monitoring of the solute distribution between adsorbed and liquid states is also a feature of the recently developed biosensor-based instruments (BIAcore, BIAlite, and IAsys), which provide a measure of the concentration of solute bound to the affinity matrix coating the biosensor surface (5). Because the basic quantitative expressions are common to all of these techniques, a brief theoretical section precedes specific considerations of the individual methods.

2. Basic theoretical expressions

In quantitative affinity chromatography there are two reactions to consider: an interaction between the solute (A) and the immobilized affinity ligand (X), and a competing reaction involving a ligand in solution (S). As indicated in *Figure 1a,* the second reaction can be in competition with X for the site(s) on the partitioning solute (A)—the situation encountered when X is a covalently immobilized form of S (e.g., in the heparin-facilitated elution of thrombin from heparin–Sepharose). Alternatively, the second reaction can give rise to competition between solute (A) and ligand (S) for affinity matrix sites (X) *(Figure 1b)*—a situation exemplified by the saccharide-mediated desorption of a glycoprotein from an immobilized lectin affinity matrix.

2.1 Characterization of the solute–matrix interaction

The first step in a quantitative affinity chromatographic study is determination of the binding constant for the solute–matrix interaction, K_{AX}. Consider, initially, the situation in which a known amount of solute (A) is introduced into a slurry of affinity matrix with accessible volume V_A*. Division of the molar amount of A by V_A* defines the total concentration of solute, $\bar{\bar{C}}_A$; although its magnitude is unknown, there is also an effective total concentration of matrix sites, $\bar{\bar{C}}_X$, whose value must evolve from the thermodynamic analysis. By measuring the concentration of A in the liquid phase, $\bar{C}_A$, we can determine the concentration of bound A, $(\bar{\bar{C}}_A - \bar{C}_A)$. A series of such experiments with a range of total solute concentrations, $\bar{\bar{C}}_A$, but the same amount and volume of slurry ($\bar{\bar{C}}_X$ constant) then generates a binding curve for the interaction of A with X. For current purposes we choose to write the rectangular hyperbolic relationship for such binding in terms of its Scatchard (6) linear transform, namely

$$(\bar{\bar{C}}_A - \bar{C}_A)/\bar{C}_A = K_{AX}\bar{\bar{C}}_X - K_{AX}(\bar{\bar{C}}_A - \bar{C}_A) \qquad [1]$$

which allows K_{AX} and $\bar{\bar{C}}_X$ to be obtained from the slope and intercept, respectively, of a plot of $(\bar{\bar{C}}_A - \bar{C}_A)/\bar{C}_A$ versus $(\bar{\bar{C}}_A - \bar{C}_A)$.

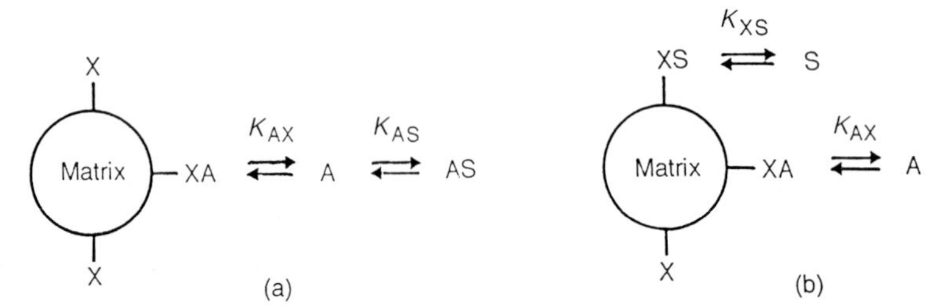

Figure 1. Schematic representation of the two competitive situations encountered in the characterization of ligand binding by quantitative affinity chromatography.

To adapt this approach to column chromatography (where $\bar{\bar{C}}_A$ is not a measurable quantity), advantage is taken of the fact that the total amount of solute, $V_A{}^*\bar{C}_A$, must equal $\bar{V}_A C_A$, the product of the concentration of A in the liquid phase and the measured elution volume, $\bar{V}_A$ (7). Substitution of $\bar{V}_A/V_A{}^*$ for $\bar{\bar{C}}_A/\bar{C}_A$ in Equation 1 then gives

$$[(\bar{V}_A/V_A{}^*) - 1] = K_{AX}\bar{\bar{C}}_X - K_{AX}\bar{\bar{C}}_A[(\bar{V}_A/V_A{}^*) - 1] \qquad [2]$$

as the column chromatographic equivalent of the Scatchard plot.

In many affinity chromatographic systems the partitioning solute is multivalent (e.g., an enzyme such as aldolase, comprising four subunits with a binding site on each). For such systems the multivalent counterpart of the Scatchard relationship (8) is

$$(\bar{\bar{C}}_A{}^{1/f} - \bar{C}_A{}^{1/f})/\bar{C}_A{}^{1/f} = K_{AX}\bar{\bar{C}}_X - fK_{AX}\bar{\bar{C}}_A{}^{(f-1)/f}(\bar{\bar{C}}_A{}^{1/f} - \bar{C}_A{}^{1/f})\,, \qquad [3]$$

where f is the valence of the partitioning solute and K_{AX} is the intrinsic binding constant (9). The chromatographic variant may then be written as

$$[(\bar{V}_A/V_A{}^*)^{1/f} - 1] = K_{AX}\bar{\bar{C}}_X - fK_{AX}(\bar{V}_A/V_A{}^*)^{(f-1)/f}\,\bar{C}_A[(\bar{V}_A/V_A{}^*)^{1/f} - 1], \quad [4]$$

whereupon K_{AX} and $\bar{\bar{C}}_X$ may be evaluated from the linear plot of $[(\bar{V}_A/V_A{}^*)^{1/f} - 1]$ versus $(\bar{V}_A/V_A{}^*)^{(f-1)/f}\,\bar{C}_A[(\bar{V}_A/V_A{}^*)^{1/f} - 1]$ after assigning a value to f, the solute valence.

2.2 Characterization of the competing reaction

In the presence of competing ligand, S, the partition behaviour of the solute clearly requires description in terms of two equilibria: the solute–matrix interaction governed by intrinsic binding constant, K_{AX}, and a second interaction governed by either K_{AS} or K_{XS} (see *Figure 1*). Nevertheless, the results obtained in the presence of a free ligand concentration, C_S, can be treated in exactly the same way as before, the only difference being that the solute–matrix equilibrium constant in Equations 1–4 is now a constitutive parameter, $\bar{K}_{AX}$, defined either as

$$\bar{K}_{AX} = K_{AX}/(1 + K_{AS}C_S) \qquad [5a]$$

or as

$$\bar{K}_{AX} = K_{AX}/(1 + K_{XS}C_S) \qquad [5b]$$

depending on which mode of competition is appropriate. The equilibrium constant for the competing reaction is thus obtained from the slope of the linear dependence of $K_{AX}/\bar{K}_{AX}$ upon C_S. Under conditions where $\bar{C}_S$, the total concentration of ligand, is the quantity available to the experimenter, the required free concentration, C_S, is obtained (4,10) from the expression

$$C_S = \bar{C}_S - (Q - 1)f\bar{C}_A/Q \qquad [6]$$

where $Q = K_{AX}/\bar{K}_{AX}$. Consequently, the same approach may be adopted by plotting Q as a function of $[\bar{C}_S - (Q - 1)f\bar{C}_A/Q]$.

3. Illustration of experimental procedures

Because of its development from preparative affinity chromatography, the initial quantitative approach also entailed the use of column chromatography to characterize the two operative equilibria for a particular system. Such techniques are therefore considered first.

3.1 Column chromatographic procedures

From Equations 2 and 4 it can be seen that the required plot for evaluating K_{AX} (or $\bar{K}_{AX}$) depends upon knowledge of $\bar{C}_A$, the total concentration of solute in the liquid phase. This quantity is not available from conventional zonal chromatography experiments because of the continual dilution that occurs during migration of the zone. Rigorous application of Equation 4 therefore requires adoption of the frontal chromatographic procedure.

3.1.1 Frontal affinity chromatography

Frontal affinity chromatography (11) differs from the zonal technique only in regard to the volume of solution applied to the column. Instead of a small sample, sufficient solution is loaded onto the column to ensure that the elution profile contains a plateau region in which the concentrations of all reactants (A, S, and solute–ligand complexes if appropriate) equal those in

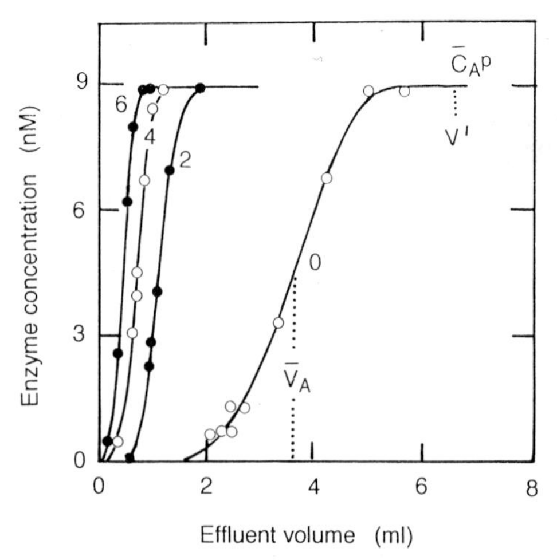

Figure 2. Effect of the indicated concentrations (μM) of NADH on the advancing elution profile obtained in frontal affinity chromatography of rat liver lactate dehydrogenase on a 0.1 ml column of 10-carboxydecylamino-Sepharose. (Adapted from ref. 12.)

the solution being applied. Because the column can be small, this requirement does not necessarily involve the use of very large volumes of solution, even though the applied volume may need to be many times larger than the column volume if the chromatographic matrix is heavily substituted with affinity groups. This aspect is illustrated in *Figure 2*, which presents elution profiles from a series of experiments on the effect of NADH concentration (C_S) in frontal chromatography of rat liver lactate dehydrogenase (9 nM) on a 0.1 ml column of 10-carboxydecylamino-Sepharose (12). In the absence of pyridine nucleotide at least 60 column volumes (6 ml) of solution were required to generate the plateau of original composition.

The above illustration of the basic tenets of frontal chromatography sets the scene for outlining the basic steps entailed in the determination of a binding constant by frontal affinity chromatography.

Protocol 1. Determination of binding constants by frontal affinity chromatography

1. Equilibrate the affinity column with buffer at the flow rate and temperature to be used for the binding study.
2. For each of a series of solutions of A with different solute concentrations, load a sufficient volume of solution to ensure the existence of a plateau region with the applied concentration ($\bar{C}_A{}^P$ in *Figure 2*).
3. Obtain the elution volume ($\bar{V}_A$) for each applied concentration as the median bisector of the elution profile, which is effectively the effluent volume at which $\bar{C}_A = \bar{C}_A{}^P/2$ if the boundary is reasonably symmetrical (13).
4. Determine the solute–matrix binding constant (K_{AX}) and the effective concentration of matrix sites ($\bar{\bar{C}}_X$) from a Scatchard plot in accordance with Equation 4, which requires prior assignment of a valence (f) to the partitioning solute (A).
5. Equilibrate the affinity column with buffer containing a known concentration, C_S, of ligand.
6. Perform frontal experiments on a range of solutions of A (with different $\bar{C}_A$) that are in dialysis equilibrium with the ligand-supplemented buffer. Note that the dialysis step to establish the free ligand concentration, C_S, may be omitted if $\bar{C}_S >> \bar{C}_A$, because $C_S \approx \bar{C}_S$ under those conditions.
7. Measure the elution volumes and evaluate the constitutive binding constant ($\bar{K}_{AX}$) from a plot of the data according to Equation 4. Note that $\bar{\bar{C}}_X$ (the abscissa intercept) must have the same magnitude as before.
8. Determine the appropriate ligand binding constant (K_{AS} or K_{XS}) from the slope of a plot of $K_{AX}/\bar{K}_{AX}$ versus C_S (see Equation 5).

Table 1. Spreadsheet for calculation of the solute–matrix binding constant from column chromatographic data on the interaction of lactate dehydrogenase with trinitrophenyl-Sepharose in the presence and absence of NADH[a,b]

	No NADH			5 μM NADH		
$\bar{C}_A$ (μM)	$\bar{V}_A$ (ml)	$[(\bar{V}_A/V_A^*)^{1/4} - 1]$	$(\bar{V}_A/V_A^*)^{3/4}\,\bar{C}_A\,[(\bar{V}_A/V_A^*)^{1/4} - 1]$ (μM)	$\bar{V}_A$ (ml)	$[(\bar{V}_A/V_A^*)^{1/4} - 1]$	$(\bar{V}_A/V_A^*)^{3/4}\,\bar{C}_A[(\bar{V}_A/V_A^*)^{1/4} - 1]$ (μM)
0.30	22.1	0.409	0.344	13.5	0.246	0.143
0.60	20.0	0.375	0.585	13.1	0.237	0.269
1.20	17.9	0.337	0.967	13.1	0.237	0.538
1.80	17.7	0.333	1.421	12.5	0.222	0.730
2.40	17.3	0.326	1.823	12.2	0.215	0.925
3.00	16.1	0.302	2.000	12.1	0.212	1.133
3.60	15.1	0.281	2.129	12.0	0.210	1.339
4.20	14.8	0.275	2.394	11.5	0.197	1.419
4.80	14.4	0.266	2.593	11.6	0.200	1.658
5.40	13.9	0.255	2.723	11.2	0.189	1.716
6.00	13.6	0.248	2.895	11.1	0.186	1.874

[a]Results obtained on a 0.9 cm x 9.5 cm column of affinity matrix, for which $V_A^* = 5.6$ ml.
[b]Results formulated for analysis in accordance with Equation 4.

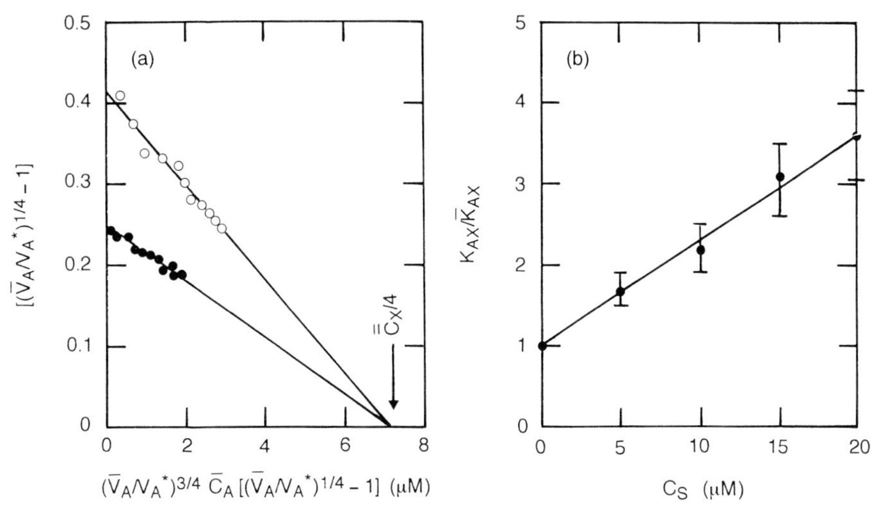

Figure 3. Characterization of the interaction between NADH and rabbit muscle lactate dehydrogenase by frontal affinity chromatography of the enzyme on trinitrophenyl-Sepharose. (a) Multivalent Scatchard plot of results *(Table 1)* obtained in the absence of coenzyme (○) and in the presence of 5 μM NADH (●). (b) Dependence of the resultant $\bar{K}_{AX}$ values upon NADH concentration, C_S (μM).

To illustrate the manipulation of data required for evaluating binding constants by this method, the relevant information obtained by frontal chromatography of rabbit muscle lactate dehydrogenase on a column (0.9 cm × 9.5 cm) of trinitrophenyl-Sepharose is summarized in *Table 1*. The first column lists the applied concentrations of enzyme ($\bar{C}_A \equiv \bar{C}_A{}^p$), and the second the elution volumes ($\bar{V}_A$) obtained for lactate dehydrogenase alone. Also presented on the right-hand side of *Table 1* is the corresponding dependence of $\bar{V}_A$ for enzyme in dialysis equilibrium with buffer containing 5 μM NADH (C_S = 5 μM). The spreadsheet in *Table 1* also shows the results of these two series of experiments in the form required for analysis in terms of Equation 4. In that regard lactate dehydrogenase is regarded as being tetravalent (f = 4) because of its tetrameric nature.

Scatchard analysis of the results is presented in *Figure 3a*, where the common abscissa intercept ($\bar{\bar{C}}_X/4$) signifies a value of 28.7 (± 0.5) μM for the total concentration of matrix sites. A binding constant of 1.4 (± 0.1) × 10^4 M^{-1} for the enzyme–matrix interaction (K_{AX}) is obtained from the slope of the plot in the absence of NADH, whereas the decreased magnitude of $\bar{K}_{AX}$ in the presence of 5 μM coenzyme reflects competition for the NADH-binding sites on the enzyme. The determination of K_{AS} for the competing interaction is summarized in *Figure 3b*, which includes results from additional series of experiments with other free NADH concentrations. From the slope of this plot in accordance with Equation 5, an intrinsic binding constant of 1.3 (± 0.1) × 10^5

M^{-1} is obtained for the interaction of NADH with rabbit muscle lactate dehydrogenase under the studied conditions (67 mM phosphate, pH 7.2).

3.1.2 Zonal affinity chromatography

In zonal chromatography the inability to ascribe a unique value to $\bar{C}_A$ because of the ever-changing solute concentration in the migrating zone clearly precludes application of the above rigorous procedure. Under those circumstances the only available option is to truncate Equation 4 as

$$[(\bar{V}_A/V_A^*)^{1/f} - 1] = K_{AX}\bar{\bar{C}}_X, \qquad [7]$$

which is a valid approximation provided that $\bar{\bar{C}}_X$ is much greater than $\bar{\bar{C}}_A$: indeed, the approximation being made is that the free concentration of matrix sites is given by $\bar{\bar{C}}_X$ (14). Subject to the acceptability of that proviso, the elution volumes ($\bar{V}_A$) obtained by applying a small zone of solute to a column of affinity matrix pre-equilibrated with a series of ligand concentrations, C_S, suffices for evaluating the competitive binding constant via the expression

$$[(\bar{V}_A/V_A^*)^{1/f} - 1] = K_{AX}\bar{\bar{C}}_X/(1 + K_{iS}C_S), \qquad [8]$$

where i = A or X; two linear transforms of this expression are

$$[(\bar{V}_A/V_A^*)^{1/f} - 1] = K_{AX}\bar{\bar{C}}_X - K_{iS}C_S[(\bar{V}_A/V_A^*)^{1/f} - 1] \qquad [9a]$$

and

$$1/[(\bar{V}_A/V_A^*)^{1/f} - 1] = 1/(K_{AX}\bar{\bar{C}}_X) + K_{iS}C_S/(K_{AX}\bar{\bar{C}}_X) \qquad [9b]$$

The former has the advantage that the required binding constant follows directly from the slope, whereas it must be obtained as the ratio of the slope and the ordinate intercept if Equation 9b is used (13). From an experimental viewpoint zonal quantitative affinity chromatography is relatively simple to perform (see *Protocol 2*).

Protocol 2. Determination of a binding constant by zonal affinity chromatography

1. Equilibrate the affinity column with buffer containing a known concentration, C_S, of ligand.
2. Subject a small zone of solute to chromatography on the column; measure the elution volume by assuming that $\bar{V}_A$ is given with sufficient accuracy by the effluent volume corresponding to the peak of the eluted zone.
3. Re-equilibrate the affinity column with buffer containing a range of concentrations of ligand; determine $\bar{V}_A$ appropriate to each C_S by the preceding zonal chromatography step.

4. Determine K_{iS} (where i = A or X) from the ($C_S,\bar{V}_A$) data set by analysis in terms of Equation 8 or either of its linear transforms (Equations 9a and 9b).

This approach is illustrated in *Figure 4* by consideration of the effect of 3′-(*p*-aminophenylphosphate)-5′-phosphate on zonal elution profiles for *Staphylococcus* nuclease on a Sepharose column to which the same ligand had been immobilized (15). Analysis of the results from *Figure 4a* on this basis in terms of Equation 9b (with i = A) is shown in *Figure 4b*, from which a value of 4.3 $\times 10^5$ M^{-1} is obtained for K_{AS} as the ratio of the slope and ordinate intercept.

4. Partition equilibrium experiments

As noted in Section 2.1, the elution volume ($\bar{V}_A$) derived from a chromatographic experiment is an indirect measure of the distribution of solute concentration between the liquid and matrix phases. We now consider alternative methods of defining this equilibrium position by determining directly the concentration of partitioning solute in the liquid phase of a solute–matrix slurry with a known total solute concentration, $\bar{\bar{C}}_A$. Two such procedures have been used.

4.1 Simple partition experiments

In its simplest form this technique has been used mainly to characterize biospecific interactions of glycolytic enzymes with cellular matrices such as

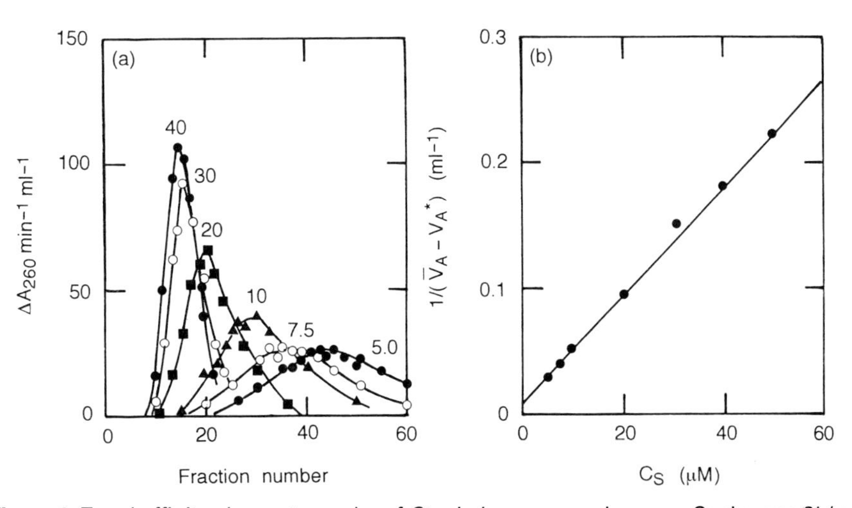

Figure 4. Zonal affinity chromatography of *Staphylococcus* nuclease on Sepharose-3′-(*p*-aminophenylphosphate)-5′-phosphate. (a) Effect of the indicated concentrations of 3′-(*p*-aminophenylphosphate)-5′-phosphate (μM) on the elution profile. (b) Evaluation of K_{AS} for the enzyme–inhibitor interaction via Equation 11b. (Adapted from ref. 15.)

muscle myofibrils (16–18) and erythrocyte membranes (19,20). Details of the procedure are as follows.

Protocol 3. Evaluation of binding constants by direct measurement of solute partition

1. Allow separate mixtures containing known amounts of affinity matrix and partitioning solute A (some without and some with ligand S) to equilibrate at the temperature of interest to establish chemical equilibrium.
2. Obtain a sample of the supernatant by filtration (3) or by centrifugation (16–20) at the temperature used for establishment of chemical equilibrium.
3. Determine the concentration of partitioning solute in the liquid phase, $\bar{C}_A$, by an appropriate spectrophotometric, radiochemical, or enzymic means: if ligand is present, $\bar{C}_A$ includes contributions of any solute–ligand complexes as well as free A.
4. Determine $\bar{\bar{C}}_A$ as the amount of added solute divided by V_A^*, the volume accessible to A in the absence of any interaction with affinity matrix.
5. Provided that all of the reaction mixtures contain the same amount of affinity matrix, plot the results in Scatchard format according to Equation 3 to obtain K_{AX}: similar treatment of data in the presence of S yields $\bar{K}_{AX}$.
6. Alternatively, accommodate a variation in the amount of affinity matrix by defining a binding function, r_f', as $(\bar{\bar{C}}_A{}^{1/f} - \bar{C}_A{}^{1/f})/\bar{\bar{c}}_X$, where $\bar{\bar{c}}_X$, (as distinct from $\bar{\bar{C}}_X$) is the total concentration of matrix sites defined on the g/litre (not the molar) scale (17).
7. Plot $r_f'/\bar{C}_A{}^{1/f}$ versus $r_f'\bar{\bar{C}}_A{}^{(f-1)/f}$ to obtain a Scatchard plot which has a slope of $-fK_{AX}$ (or $-f\bar{K}_{AX}$ in the presence of S) and an intercept corresponding to the matrix capacity for solute (mol/g) divided by f, the solute valence.

This approach is illustrated by considering published results (16) for the interaction of aldolase with rabbit muscle myofibrils. *Table 2* presents the spreadsheet for calculating the binding function (r_f') for interaction of this tetravalent solute ($f = 4$) with the myofibrillar matrix; *Figure 5* shows the analysis of the results in accordance with step 7 of *Protocol 3*. An intrinsic binding constant of 4.0 ($\pm$ 0.1) $\times$ 10^5 M^{-1} for the aldolase–myofibrils interaction is obtained from the slope ($-4K_{AX}$) of this multivalent Scatchard plot, the abscissa intercept of which signifies a matrix capacity of 70 ($\pm$ 3) nmol aldolase per gram of myofibillar protein.

Table 2. Spreadsheet for evaluation of the solute–matrix binding constant (K_{AX}) from partition equilibrium data on the interaction of aldolase with muscle myofibrils[a]

$\bar{\bar{c}}_X$ (mg/ml)	$\bar{\bar{C}}_A$ (μM)	$\bar{C}_A$ (μM)	$10^4 r_f'$	$r_f'/\bar{C}_A^{1/4}$ (ml/mg)	$r_f'\bar{\bar{C}}_A^{3/4}$ (nmol/g)
21.9	0.55	0.16	3.30	0.0165	6.66
27.9	0.68	0.15	3.24	0.0165	7.67
27.3	0.68	0.16	3.19	0.0160	7.55
21.6	1.00	0.39	3.07	0.0123	9.71
28.2	1.27	0.44	2.77	0.0108	10.48
29.7	1.35	0.44	2.78	0.0108	10.95
22.8	1.99	1.07	2.37	0.0074	12.56
28.0	2.13	1.00	2.35	0.0074	13.10
22.6	2.97	1.86	2.03	0.0055	14.52
30.9	3.26	1.79	1.91	0.0052	14.65
21.8	3.95	2.79	1.70	0.0042	15.06
21.7	4.91	3.73	1.44	0.0033	15.02
29.4	5.73	4.13	1.31	0.0029	15.34
22.2	6.10	4.81	1.29	0.0028	15.83
22.8	7.48	6.13	1.11	0.0022	15.88

[a]Data taken from Table 1 of ref. 16.

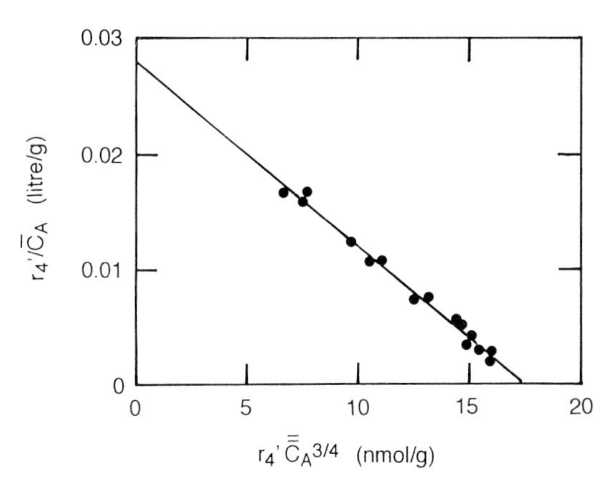

Figure 5. Multivalent Scatchard analysis (Equation 12) of partition equilibrium data (*Table 2*) for the interaction of aldolase with rabbit muscle myofibrils.

4.2 Recycling partition technique

A disadvantage of the simple partition experiment is the need for precise control of the amount of affinity matrix included in each reaction mixture—a difficult task in instances where the affinity matrix is being dispensed as a concentrated slurry. Provided that the concentration of partitioning solute is amenable to continuous assessment by means of a flow-cell monitor, this dif-

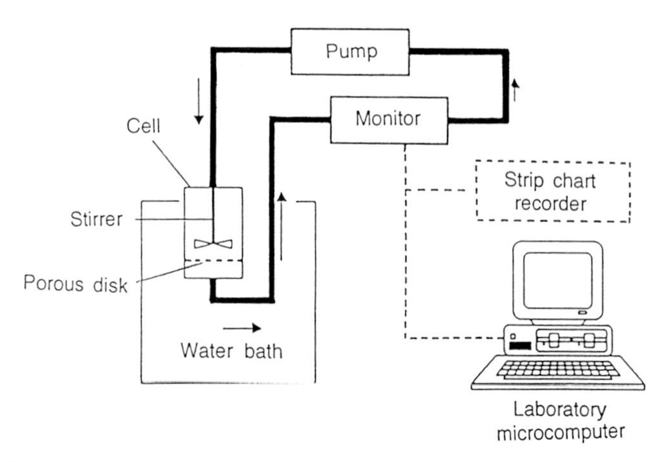

Figure 6. Schematic representation of the recycling partition equilibrium apparatus. (Adapted from ref. 4.)

ficulty can be avoided (4,21) by using a recycling partition technique which overcomes the need for any assumptions about identity of matrix amounts in a series of equilibrating mixtures. The apparatus comprises a 10 ml Amicon ultrafiltration cell (devoid of membrane) positioned over a submersible magnetic stirrer in a water bath that is thermostatically maintained at the required temperature (*Figure 6*). Effluent from below the porous membrane support is returned to the slurry via a peristaltic pump and the flow-cell of a monitor operating at 280 nm or some other suitable wavelength. The monitor response is relayed to a recorder or, if available, a data acquisition system. Procedural details are given in *Protocol 4*.

Protocol 4. A recycling variant of the partition equilibrium method of determining binding constants

1. Place a slurry (approximately 6 ml) of affinity matrix in the cell. Determine the combined weight of the cell and the slurry so that an accurate record of slurry volume in the binding experiment can be maintained (see step 3).
2. Determine V_A^*, the volume accessible to the partitioning solute (A), by equilibrating the slurry with a medium that completely suppresses the solute–matrix interaction (e.g., buffer supplemented with an extremely high concentration of a competing ligand). Add successive aliquots (10–20 μl) of a stock solution of A; use the resultant concentration of solute in the liquid phase after each addition ($\bar{C}_A$) to calculate V_A^* on the basis that $V_A^*\bar{C}_A$ is the amount of solute added.
3. Remove the liquid phase and wash the affinity matrix copiously with buffer in readiness for partition studies of the solute–matrix inter-

action. Calculate $(V_A{}^*)_o$, the accessible volume prior to solute addition, on the basis of the weight of the equilibrated cell in relation to that for the calibration experiment: $(V_A{}^*)_o = V_A{}^* + \delta w/\rho$, where δw is the difference in weights and ρ is the buffer density.

4. Add an aliquot (10 or 20 μl) of stock solute solution to the slurry, and allow the interaction to proceed until attainment of equilibrium is indicated by constancy of the monitor response.
5. Repeat this procedure with further aliquots of stock solute solution to generate sufficient data for characterizing the solute–matrix interaction via a Scatchard plot in accordance with Equation 3.
6. Now add successive aliquots of a stock solution of the soluble ligand (S) to provide results for the evaluation of K_{AS} or K_{XS}, whichever is the appropriate competing interaction (see *Figure 1*).

The procedure described in *Protocol 4* is illustrated by considering the results of an investigation in which heparin–Sepharose has been used as the affinity chromatographic matrix to characterize the interaction between heparin and antithrombin (4). *Table 3* summarizes the basic results of an experiment in which aliquots of antithrombin solution (0.28 mM) were first added to the stirred slurry of heparin–Sepharose: in the second stage the slurry was supplemented with aliquots of high-affinity heparin (1.05 mM). The following points are noted.

(a) The first column lists the accessible volume of the liquid phase, $V_A{}^*$,

Table 3 Spreadsheet for calculating K_{AX} and $\bar{K}_{AX}$ from a recycling partition study of the effect of high-affinity heparin (S) on the interaction between antithrombin (A) and heparin–Sepharose (X)[a]

Volume added (μl)[b] A	S	$V_A{}^*$ (ml)	$\bar{\bar{C}}_A$ (μM)	$\bar{C}_S$ (μM)	$\bar{C}_A$ (μM)	$\bar{\bar{C}}_X$ (μM)	$\bar{K}_{AX}$ (μM^{-1})
–	–	5.96	–	–	–	–	–
20	–	5.98	0.937	–	0.094	–	–
10	–	5.99	1.404	–	0.177	–	–
10	–	6.00	1.869	–	0.349	–	–
10	–	6.01	2.332	–	0.599	–	–
10	–	6.02	2.794	–	0.908	–	–
10	–	6.03	3.254	–	1.423	–	–
10	–	6.04	3.713	–	1.881	–	–
–	20	6.06	3.700	3.465	2.937	2.164	0.1855
–	10	6.07	3.694	5.189	3.225	2.160	0.0944
–	10	6.08	3.688	6.908	3.357	2.157	0.0540
–	10	6.09	3.682	8.621	3.439	2.153	0.0370

[a]Data taken from Table 2 of ref. 4.
[b]Concentrations of stock solutions: A, 0.28 mM; S, 1.05 mM.

which increases progressively with the addition of aliquots of antithrombin and, subsequently, heparin.

(b) Values of $\bar{\bar{C}}_A$ are calculated as the amount of added antithrombin divided by the relevant V_A*, whereas those of $\bar{C}_A$ are based on the monitored equilibrium absorbance at 280 nm of the liquid phase.

(c) The magnitude of K_{AX} is then determined by means of Equation 3 with $f = 1$ for this interaction between univalent antithrombin and the heparin–Sepharose affinity matrix. In so doing, allowance must be made for the decrease in $\bar{\bar{C}}_X$ that results from the increase in V_A*. Inspection of the Scatchard plot (*Figure 7a*) shows that this dilution has been taken into account by multiplying the conventional ordinate and abscissa parameters by $V_A^*/(V_A^*)_o$, whereupon the quantity defined by the abscissa intercept is $(\bar{\bar{C}}_X)_o$, the effective total concentration of matrix sites at the commencement of solute addition. Values of 6.8 (± 0.3) × 10^6 M^{-1} and 2.2 (± 0.4) μM are obtained for K_{AX} and $(\bar{\bar{C}}_X)_o$ respectively.

(d) Knowledge of $(\bar{\bar{C}}_X)_o$ now allows magnitudes to be assigned to $\bar{\bar{C}}_X$ at each stage of the second part of the experiment in which soluble heparin was added to displace antithrombin from the affinity matrix. Consequently, $\bar{K}_{AX}$ is calculated as the only parameter of unknown magnitude in Equation 3, which is written more conveniently in the form

$$\bar{K}_{AX} = (\bar{\bar{C}}_A^{\,1/f} - \bar{C}_A^{\,1/f})/\{\bar{C}_A^{\,1/f}[\bar{\bar{C}}_X - fK_{AX}\bar{\bar{C}}_A^{\,(f-1)/f}(\bar{\bar{C}}_A^{\,1/f} - \bar{C}_A^{\,1/f})]\} \quad [10]$$

The resulting values of $\bar{K}_{AX}$ are presented in the final column of *Table 3*.

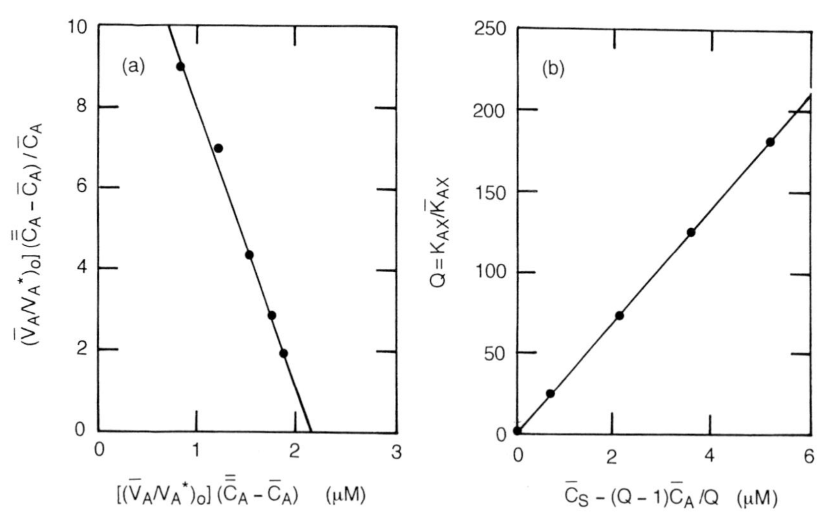

Figure 7. Characterization of ligand binding by the recycling partition technique of quantitative affinity chromatography. (a) Scatchard plot of results (*Table 3*) for the interaction of antithrombin with heparin–Sepharose. (b) Evaluation of the binding contant (K_{AS}) for the interaction of antithrombin (A) with high-affinity heparin (S) from results also presented in *Table 3*.

(e) Because $\bar{C}_S$, the total concentration of heparin, is the quantity deduced from the amount added, Equation 6 must be used to define the magnitude of C_S (the free ligand concentration) that is required for evaluating K_{AS} via Equation 5a. From the consequent plot (*Figure 7b*), a binding constant of 3.4 ($\pm$ 0.3) $\times$ 10^7 M^{-1} is obtained for the interaction of antithrombin with high-affinity heparin.

5. Biosensor-based procedures

An alternative way of quantifying solute partitioning between adsorbed and liquid states is to measure the concentration of bound solute in the equilibrium mixture by the recently developed biosensor technologies incorporated into the BIAcore or BIAlite (Pharmacia) and IAsys (Affinity sensors) instruments. An affinity chromatographic matrix is prepared *in situ* by covalently attaching the appropriate ligand to a carboxymethylated dextran layer on the biosensor surface (a gold chip in the Pharmacia instruments, and the cuvette base in IAsys). Different biosensor technologies are employed in the two instruments (22,23), but they both exhibit similar sensitivities of response to bound solute. Because the mass of bound solute is the parameter actually being monitored, the methodology is far more sensitive in molar terms for macromolecules than for small partitioning solutes: for proteins the optimal range of bound concentration is 5–100 nM. Thus far the major interest has been their use to define the kinetics of association and dissociation for the solute–matrix interaction on the basis that the partitioning solute is univalent in its interaction with matrix sites (24–26). Such neglect of the possibility that a macromolecular partitioning solute may well exhibit multivalence in its interaction with matrix sites is clearly a deficiency of this approach. Attention is therefore directed towards use of the biosensor technology for thermodynamic characterization of the solute–matrix interaction as an initial step in the determination of a binding constant for a competing reaction between solute and ligand in solution (5). Because the manner in which the sample is presented to the biosensor differs in the two types of commercial instrument, slightly different analytical procedures need to be implemented: that for the BIAcore (and BIAlite) is described first.

5.1 Binding constants from BIAcore (BIAlite) responses

In the BIAcore and BIAlite instruments, which differ only in regard to the degree of automation, the solution of partitioning solute that is injected flows through a capillary channel over the biosensor—a gold chip with an area of 1.1 mm^2: the dimensions of the channel are such that the volume of solution in the immediate vicinity of the chip is 0.06 µl. Injection of a much larger volume of solute solution (50 µl) thus gives rise to a response of the form shown in *Figure 8*. Initially there is a rapid increase in biosensor response, reflecting

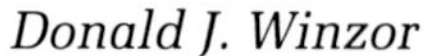

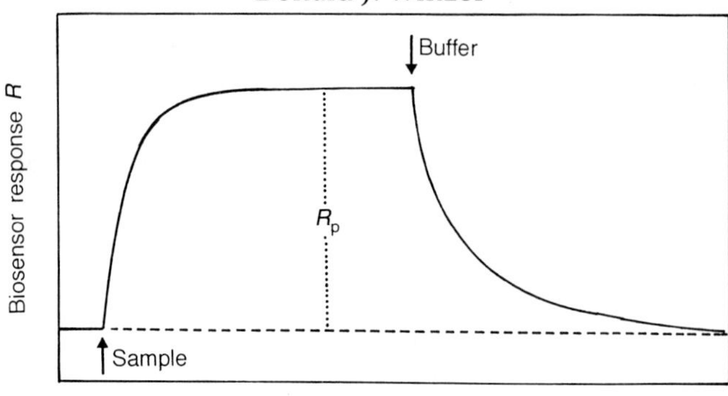

Figure 8. Schematic representation of the time-dependence of biosensor response resulting from injection of a relatively large volume (50 μl) of partitioning solute into the BIAcore (or BIAlite) instrument.

the kinetics of the interaction between matrix sites and partitioning solute. A time-dependent response thus signifies a lack of equilibrium between the concentration of bound solute and that in the solution with which the bound solute is in contact. However, provided that a sufficient volume of solution passes across the sensor chip, a stage is reached where the biosensor response becomes time-independent. Correction of this plateau response for that prior to injecting the solution (the baseline response) gives the parameter R_p (*Figure 8*), which is directly proportional to the concentration of bound solute, $(\bar{\bar{C}}_A - \bar{C}_A)$, in equilibrium with $\bar{C}_A$, the concentration of partitioning solute in the injected solution.

5.1.1 Analytical procedure for a univalent solute

For a partitioning solute that is univalent in its interaction with immobilized affinity sites on the biosensor chip, the measurement of R_p as a function of the injected solute concentration, $\bar{C}_A$, suffices to determine K_{AX}, because the Scatchard formulation for this system may be written as

$$R_p/\bar{C}_A = \mathrm{B}K_{AX}\bar{\bar{C}}_X - K_{AX}R_p \,, \qquad [11]$$

where B is the proportionality constant relating the corrected plateau response to bound concentration $[R_p = \mathrm{B}(\bar{\bar{C}}_A - \bar{C}_A)]$. Experimental details of the thermodynamic approach to use of the BIAcore (BIAlite) technology are given in Protocol 5.

Protocol 5. Use of BIAcore technology to determine binding constants for a univalent solute

1. Equilibrate the surface of the biosensor with the buffer to be used in the binding study so that the baseline response (see *Figure 8*) is accurately defined.

2. Inject successive aliquots (50 μl) of solutions of A with different concentrations and record the plateau response for each solution (with concentration $\bar{C}_A$).
3. Determine K_{AX} and $B\bar{\bar{C}}_X$ from the slope and abscissa intercept respectively of the Scatchard plot of $R_p/\bar{C}_A$ versus R_p (Equation 11).
4. Inject aliquots (50 μl) of solutions of A supplemented with a range of known concentrations of competing ligand, $\bar{C}_S$.
5. For each experimental $[R_p, \bar{C}_A, \bar{C}_S]$ point, calculate the constitutive binding constant as $\bar{K}_{AX} = R_p/\{\bar{C}_A[B\bar{\bar{C}}_X - K_{AX}R_p]\}$ on the basis of the values of $B\bar{\bar{C}}_X$ and K_{AX} determined in step 3.
6. Calculate $Q = K_{AX}/\bar{K}_{AX}$ for the required plot of Q versus $[\bar{C}_S - (Q - 1)\bar{C}_A/Q]$ to obtain K_{AS}, as in *Figure 7b* for the analogous analysis of recycling partition data.

Use of BIAcore technology to evaluate the solute–matrix binding constant for a univalent partitioning solute is illustrated in *Figure 9a* for the interaction between immobilized interleukin-6 and the soluble form of its biospecific receptor (5). The slope signifies an intrinsic binding constant (K_{AX}) of 2.4 (± 0.2) × 10^7 M^{-1}, whereas the abscissa intercept implies a binding capacity equivalent to 2200 (± 100) response units.

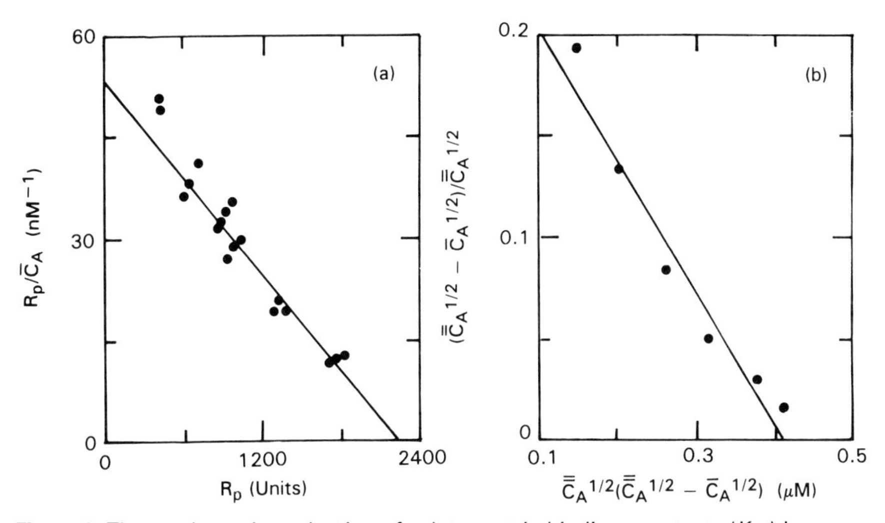

Figure 9. Thermodynamic evaluation of solute–matrix binding constants (K_{AX}) by means of biosensor-based technology. (a) Analysis of results (5) for the binding of soluble interleukin-6 receptor to a biosensor chip with interleukin-6 as the immobilized ligand. (b) Corresponding analysis of results (*Table 4*) for the interaction of concanavalin A, a bivalent solute, with the carboxymethylated dextran on the biosensor chip.

5.1.2 Allowance for multivalence of the solute

In general, the above procedure cannot be applied if the partitioning solute (A) is multivalent in its interaction with immobilized affinity sites on the biosensor chip. Although the equilibrium response (R_p) continues to describe the product $B(\bar{\bar{C}}_A - \bar{C}_A)$, this is not directly proportional to the parameter required for the application of Equation 3, namely $(\bar{\bar{C}}_A{}^{1/f} - \bar{C}_A{}^{1/f})$, unless $f = 1$. In situations where f is greater than unity, the calculation of this difference between concentrations raised to the appropriate power requires knowledge of $\bar{\bar{C}}_A$, which includes the solute bound to affinity sites on the biosensor chip. For proteins this total concentration of partitioning solute is calculated from $\bar{C}_A$, the concentration of A in the injected solution, via the expression

$$\bar{\bar{C}}_A = \bar{C}_A + [R_p/(60\,000 \times M_A) \,, \qquad [12]$$

in which the proportionality constant (B of Equation 11) has been calculated (27) from the linear calibration plot relating R_p to the surface concentration of protein on the biosensor matrix (28): M_A, the molar mass of solute, is introduced to convert the concentration of bound solute from a weight to a molar basis.

Use of biosensor technology to characterize the solute–matrix interaction for a multivalent solute is summarized in *Table 4*, which includes the requisite data from a BIAcore study (27) of the uptake of concanavalin A, a bivalent solute, by the carboxymethyldextran affinity matrix on the biosensor chip. The plot of these results in accordance with Equation 3 (*Figure 9b*) signifies an intrinsic binding constant (K_{AX}) of 3.4 ($\pm$ 0.8) $\times$ 10^5 M^{-1} for the interaction of the lectin with the immobilized carboxymethyldextran on the biosensor chip, which has an effective total capacity for concanavalin A ($\bar{\bar{C}}_X$) of 0.81 ($\pm$ 0.16) μM (twice the abscissa intercept of *Figure 9b*).

The small magnitude of $\bar{\bar{C}}_X$ relative to $\bar{C}_A$ draws attention to the sensitivity of the biosensor response for quantifying the extent of protein binding to the matrix-coated chip. Indeed, for the particular system studied in *Figure 9b*,

Table 4 Evaluation of the total concentration of concanavalin A ($\bar{\bar{C}}_A$) from the injected concentration ($\bar{C}_A$) and the equilibrium BIAcore response[a]

$\bar{C}_A$ (μM)	R_p (units)	$(\bar{\bar{C}}_A - \bar{C}_A)$ (μM)[b]	$\bar{\bar{C}}_A$ (μM)
25.00	2544	0.82	25.82
12.50	2387	0.76	13.26
6.25	2010	0.64	6.89
3.13	1672	0.54	3.67
1.56	1370	0.44	2.00
0.78	1030	0.33	1.11

[a]Data taken from ref. 27.
[b]Calculated as $R_p/(60\,000 \times M_A)$: see Equation 16.

this extreme sensitivity of the BIAcore detector allows a simplifying approximation to be used in evaluating K_{AX}. Provided that $(\bar{\bar{C}}_A - \bar{C}_A)$ is very much less than $\bar{C}_A$ (as in *Table 4*), biosensor results may be subjected to conventional Scatchard analysis (Equation 11) despite multivalence of the partitioning solute (27). However, the slope of the plot in such circumstances yields the product fK_{AX}.

5.2 Consideration of the IAsys system

In the IAsys instrument the affinity matrix is prepared by covalent attachment of a biospecific ligand to a layer of carboxymethylated dextran attached to the base of the cuvette, which is the source of the biosensor response. Whereas flow through a channel with extremely small cross-sectional area is used to achieve uniformity of the liquid phase in the BIAcore instrument, the contents of the IAsys cuvette are stirred to achieve the same effect. The time-course of the biosensor response assumes the same form as that observed with the BIAcore and BIAlite instruments (see *Figure 8*), but in this instance the plateau response reflects a concentration of bound solute in equilibrium with a liquid-phase concentration, $\bar{C}_A$, that needs to be determined from $\bar{\bar{C}}_A$, the total concentration inferred from the amount of solute added, and the biosensor response. Thus, even for a univalent solute, the magnitude of the calibration constant (B) is required to convert the biosensor response to a concentration difference so that $\bar{C}_A$ may be determined from $\bar{\bar{C}}_A$ and R_p by the IAsys analogue of Equation 12, namely

$$\bar{C}_A = \bar{\bar{C}}_A - (100/V)[R_p/(815\,000 \text{ x } M_A)] \qquad [13]$$

where the biosensor response is recorded in arc sec, and V is the volume (in microlitres) of solution in the cell. The additional term in Equation 13 accommodates the difference between the actual solution volume in the cuvette and the value of 100 μl used to calculate B from a calibration plot in terms of surface concentration.

This consideration of the IAsys instrumentation for rigorous thermodynamic characterization of ligand binding is premature inasmuch as its use for such purposes is yet to be reported. However, such studies will undoubtedly be forthcoming, because the instrumental design is ideal for adoption of a stepwise titration procedure akin to that for determining binding constants by the recycling partition technique (see *Protocol 4*). Details of the method are given in *Protocol 6*.

Protocol 6. Potential use of IAsys technology to determine binding constants

1. Place a known volume, $(V_A{}^*)_o$, of buffer (50–200 μl) in the cuvette in order to establish a baseline response (see *Figure 8*).
2. Add an aliquot, δV (10 or 20 μl), of a stock solution of A to the cuvette,

Protocol 6. *Continued*

and allow the interaction to proceed until attainment of equilibrium is indicated by constancy of the biosensor response.

3. Calculate $\bar{C}_A$ from Equation 13 on the basis of the plateau response (R_p) and the magnitude of $\bar{\bar{C}}_A$, which is obtained by dividing the amount of solute added by $[(V_A^*)_o + \delta V]$.
4. Repeat steps 2 and 3 with further aliquots of the stock solution of A to generate sufficient data for characterizing the solute–matrix interaction via a Scatchard plot in accordance with Equation 3, or Equation 11 if A is univalent. In that regard the incremental increases in V_A^* with each aliquot addition necessitate multiplication of the conventional ordinate and abscissa parameters by $V_A^*/(V_A^*)_o$, as in *Figure 7*.
5. Now add successive aliquots (10 or 20 μl) of a stock solution of ligand (S) to provide results for the evaluation of K_{AS} or K_{XS}, whichever describes the competing interaction (see *Figure 1*). Refer to *Table 3* for a spreadsheet describing the analogous analysis of recycling partition results.

5.3 Relative merits of the two technologies

Comparisons must obviously be tentative in the present situation where only one of these two technologies has actually been used to determine the binding constant for a competing ligand interaction. However, in principle there are several instances where one technology has advantages over the other.

(a) The need for conversion of R_p to a concentration difference in order to determine $\bar{C}_A$ for univalent as well as multivalent partitioning solutes (unless $\bar{\bar{C}}_A$ can justifiably be substituted for $\bar{C}_A$) is a disadvantage of the IAsys instrument.

(b) The IAsys instrumental design should be advantageous if partition equilibrium is attained very slowly. Whereas slow attainment of equilibrium in the BIAcore and BIAlite instruments may well require the injection of rather large volumes of solution (500–1000 μl), a longer time requirement for attainment of equilibrium in the IAsys system merely tests the instrument's stability of response and the experimenter's patience.

(c) An advantage of the BIAcore (or BIAlite) instrumental design is the unequivocal identification of the equilibrium concentration of partitioning solute in the liquid phase ($\bar{C}_A$) as that of the injected solution: $\bar{C}_A$ has to be determined from R_p and $\bar{\bar{C}}_A$ in the IAsys system. This feature could be important in studies of very tight binding (where K_{AX} is extremely large), because $\bar{C}_A$ may be much smaller than $(\bar{\bar{C}}_A - \bar{C}_A)$, in which case the IAsys requirement that $\bar{C}_A$ be calculated as $\bar{\bar{C}}_A - (\bar{\bar{C}}_A - \bar{C}_A)$ may well lead to a value indistinguishable from zero.

(d) The ability of the IAsys instrumental design to accommodate sequential additions of partitioning solute and ligand places this set-up at a decided advantage over the BIAcore (BIAlite) system from the viewpoint of amounts of materials required to perform the stepwise titrations of the biosensor affinity matrix with partitioning solute, first alone and then in the presence of competing ligand.

In summary, the use of biosensor technology (BIAcore, BIAlite, or IAsys) to characterize ligand binding is still in its relative infancy, and much more work needs to be done in order to explore fully the potential advantages and limitations of the approach. Nevertheless, the results obtained thus far are sufficiently encouraging to signify that biosensor technology will play an important role in the thermodynamic characterization of biospecific interactions, particularly those equilibria governed by very large association constants.

6. Concluding remarks

A major advantage of quantitative affinity chromatography for the characterization of ligand binding is its versatility. Because the basic quantitative expressions can be applied to column chromatographic data as well as to results obtained by a variety of partition equilibrium procedures, a researcher can choose the experimental technique most able to accommodate any practical limitations asociated with study of a particular interaction (amounts and stability of reactants, etc.). Quantitative affinity chromatography is also a versatile technique in the sense that there is no restriction on the size of the ligand: it has been used to characterize the binding of small molecules to macromolecules (1–3,7,10,14,15,21), of macromolecules to macromolecules (4,5,32), and of macromolecules to subcellular assemblies (16–20). Possibly the greatest asset of quantitative affinity chromatography from the viewpoint of versatility is its ability to characterize an extraordinarily wide range of binding affinities: the method has been used to characterize relatively weak ($K_{AS} < 10^3\ M^{-1}$) interactions (21) as well as very strong ($K_{AS} > 10^9\ M^{-1}$) ones (29). Finally, if the specificity exhibited by the affinity matrix for a solute is absolute, quantitative affinity chromatography has the potential to characterize interactions of solutes (A) that have not been purified to homogeneity—a feature exemplified by the determination of equilibrium constants for the binding of NADH to the various isoenzymes of lactate dehydrogenase present in a crude mouse tissue extract (30).

Note added in proof: There have been several advances in the characterization of ligand binding by biosensor technology since the submission of this manuscript. In particular, it should be noted that the envisaged IAsys developments have now been reported [Hall, D. R. and Winzor, D. J. (1997) *Anal. Biochem.* **244**, 152–60].

References

1. Andrews, P., Kitchen, B. J., and Winzor, D. J. (1973). *Biochem. J.*, **135**, 897.
2. Dunn, B. M. and Chaiken, I. M. (1974). *Proc. Natl Acad. Sci. USA*, **71**, 2382.
3. Nichol, L. W., Ogston, A. G., Winzor, D. J., and Sawyer, W. H. (1974). *Biochem. J.*, **143**, 435.
4. Hogg, P. J., Jackson, C. M., and Winzor, D. J. (1991). *Anal. Biochem.*, **192**, 303.
5. Ward, L. D., Howlett, G. J., Hammacher, A., Weinstock, J., Yasukawa, K., Simpson, R. J., and Winzor, D. J. (1995). *Biochemistry*, **34**, 2901.
6. Scatchard, G. (1949). *Ann. N.Y. Acad. Sci.*, **51**, 660.
7. Hogg, P. J. and Winzor, D. J. (1984). *Arch. Biochem. Biophys.*, **234**, 55.
8. Hogg, P. J. and Winzor, D. J. (1985). *Biochim. Biophys. Acta*, **843**, 159.
9. Klotz, I.M. (1946). *Arch. Biochem.*, **9**, 109.
10. Winzor, D. J., Munro, P. D., and Jackson, C. M. (1992). *J. Chromatogr.*, **597**, 57.
11. Winzor, D. J. and Scheraga, H. A. (1963) *Biochemistry*, **2**, 1263.
12. Kyprianou, P. and Yon, R. J. (1982). *Biochem. J.*, **207**, 549.
13. Winzor, D. J. and Jackson, C. M. (1993). In *Handbook of affinity chromatography* (ed. T. Kline), p. 253. Marcel Dekker, New York.
14. Bergman, D. A. and Winzor, D. J. (1986) *Anal. Biochem.*, **153**, 380.
15. Dunn, B. M. and Chaiken, I. M. (1975). *Biochemistry*, **14**, 2343.
16. Kuter, M. R., Masters, C. J., and Winzor, D. J. (1983) *Arch. Biochem. Biophys.*, **225**, 384.
17. Harris, S. J. and Winzor, D. J. (1989). *Arch. Biochem. Biophys.*, **275**, 185.
18. Harris, S. J. and Winzor, D. J. (1989). *Biochim. Biophys. Acta*, **999**, 95.
19. Kelley, G. E. and Winzor, D. J. (1984). *Biochim. Biophys. Acta*, **778**, 67.
20. Harris, S. J. and Winzor, D. J. (1990). *Biochim. Biophys. Acta*, **1038**, 306.
21. Nichol, L. W., Ward, L. D., and Winzor, D. J. (1981). *Biochemistry*, **20**, 4856.
22. Löfås, S. and Johnsson, B. (1990). *J. Chem. Soc. Chem. Commun.*, 1526–1528.
23. Cush, R., Cronin, J. M., Stewart, W. J., Maule, C. H., Molloy, J., and Goddard, N. J. (1993). *Biosensors & Bioelectronics*, **8**, 347.
24. Karlsson, R., Michaelson, R., and Mattson, L. (1991). *J. Immunol. Methods*, **145**, 229.
25. Zeder-Lutz, G., Altschuh, D., Geysen, H. M., Trifilieff, E., Sommermayer, G., and Van Regenmortel, M. H. V. (1993). *Mol. Immunol.*, **30**, 145.
26. O'Shannessy, D. J., Brigham-Burke, M., Soneson, K. K., Hensley, P., and Brooks, I. (1993). *Anal. Biochem.*, **212**, 457.
27. Kalinin, N. L., Ward, L. D., and Winzor, D. J. (1995). *Anal. Biochem.*, **228**, 238.
28. Stenberg, E., Persson, B., Roos, H., and Urbaniszky, V. (1991). *J. Colloid Interface Sci.*, **143**, 513.
29. Waltham, M. C., Holland, J. W., Nixon, P. F., and Winzor, D. J. (1988) *Biochem. Pharmacol.*, **37**, 541.
30. Brinkworth, R. I., Masters, C. J., and Winzor, D. J. (1975) *Biochem. J.*, **151**, 631.

3

Affinity separations of proteins

PAUL MATEJTSCHUK, PETER A. FELDMAN, JACKY ROTT, and JOHN MORE

1. Introduction

Probably the greatest proportion of affinity-based separations reported in the scientific literature involve proteins. There are literally thousands of references to the isolation, characterization, and utilization of proteins that rely upon affinity-based interactions for successful separations (1). This being the case it would be impossible to cover adequately here all of the methodologies documented, nor is it the purpose of this chapter merely to review the application of affinity chromatography to protein chemistry. In this chapter we will present the common techniques used for the affinity extraction of proteins and give examples of these from our own experience, such that the general principles of each group of separations can be outlined for readers to follow and apply to their own experimental situations.

Advances in, and selection of, conjugation chemistries, choice of matrix or indeed the format in which protein separations can take place are covered in Chapter 1. One growing area of affinity purification of proteins is the use of affinity tags to facilitate protein isolation; this is covered in Chapter 7. Another particularly important area of protein affinity separations is that based upon the use of an immunoglobulin molecule or fragment for the immunoaffinity purification of proteins. This, together with the use of immunoglobulin-binding receptor molecules (e.g. Proteins A and G) for antibody purification, are addressed in Chapter 6. In this chapter, we describe each of the main protein affinity separation techniques and illustrate how they have been used in the laboratory.

Many of our examples come from the application of affinity chromatography for the isolation of proteins from plasma, where most of our research interests are centred. Plasma is an extraordinarily rich source of proteins with over 100 known protein components and concentrations varying from nanograms to milligrams per millilitre and, as such, provides innumerable illustrations of successful separation strategies based upon selective molecular interaction. However, the protocols presented should not be seen to be limited to this source material as they could be applied equally well to other biotechnological protein mixtures where individual components must be isolated.

2. Affinity chromatography on immobilized protein ligands

2.1 Introduction

This section concentrates on the exploitation of protein–protein interactions to achieve affinity separations of targeted protein molecules. When considering the use of protein ligands it is important to consider the effect of the chosen conjugation chemistry on the interaction between the protein to be immobilized and the target for purification. If the chemistry of the conjugation results in the modification of residues (e.g. amino groups) important in the protein–protein interaction then the specificity of that interaction may be reduced or the entire interaction abolished. Similarly, if the orientation of the protein ligand, following conjugation to the matrix, is sterically restricted so as to prevent the adoption of the necessary stereochemistry for the interaction with target protein to occur, this too will limit the usefulness of the resin for purification of the target protein.

It is important also to consider the effects of any cleaning regime on the immobilized protein. Harsh denaturants will result in unfolding of the immobilized protein. If this occurs only partial refolding may be possible once the denaturant is removed, or indeed all functional activity may be lost. The most useful protein ligands are therefore unlikely to be those which consist of multiple chains. The most suitable proteins for immobilization are those which are highly resistant to denaturation or which refold readily to the native conformation. For example, *Staphylococcus aureus* Protein A and *Streptococcus* Protein G can recover full binding capacity after treatment with 6 M guanidinium chloride.

Examples of some of the commercially available protein ligands are given in *Table 1*.

2.2 Purification of fibronectin from human plasma on immobilized gelatin

Fibronectin is an adhesive protein present at concentrations of ~200 μg/ml in human plasma. Affinity chromatography, based on the interaction between a specific domain of fibronectin which possesses gelatin-binding activity and immobilized gelatin, is the most common method of isolation (2). Gelatin is a degraded form of collagen.

Commercially available examples of suitable media include gelatin–Sepharose CL-4B (Pharmacia) and gelatin–Ultrogel (BioSepra). Alternatively, gelatin can be coupled to suitable media such as agarose (e.g. Sepharose 4B or Sepharose CL-4B) using *Protocol 1*.

Table 1. Commercially available immobilised protein resins for affinity chromatography[a]

Immobilised protein	Typical suppliers	Matrix	Uses
Protein A/G	Pharmacia, Bioprocessing Ltd., Pierce, Perspective, ACL,	agarose, controlled pore glass, silica, synthetic polymer	immunoglobulin purification
Avidin/streptavidin	Pharmacia, Sigma ACL, Pierce, Biosepra	agarose, acrylic	purification of biotinylated proteins
Calmodulin	ACL, Pharmacia, Biosepra	agarose	purificatioin of calcium-binding proteins
Gelatin	Pharmacia, Bioprocessing Ltd, Pierce, Amicon, Biosepra	agarose, controlled pore glass, cellulose	purification of fibronectin
IgG	Pharmacia	agarose	purification of Protein A/G or proteins containing a Protein A/G affinity tag
Fetuin	Sigma	agarose	purification of haem
Haemoglobin	Sigma	agarose	purification of wheat protease
α-Lactalbumin	Sigma	agarose	purification of galactosyl transferases
Albumin	Sigma	agarose	purification of haem peptides of cytochrome *c*
Jacalin	Pharmacia, Pierce, Sigma	agarose	purificatioin of IgA_1
Thyroglobulin	Sigma	agarose	isolation of *P. vulgaris* agglutinin
Casein	Sigma	agarose	isolation of casein kinase
Cytochrome *c*	Sigma	agarose	isolation of cytochrome oxidase
Insulin	Sigma	agarose	binding to insulin-specific protease
Mannose binding protein	Pierce	agarose	purification of IgM
Collagen	Sigma	agarose	purification of collagenase
Histone	Sigma	agarose	purification of protein kinases
Aprotinin	Supelco	agarose	puriification of typsin
Concanavalin A	Pharmacia, Sigma, Pierce Bioprocessing Ltd	agarose, controlled pore glass	purification of glycoproteins
Anhydrotrypsin	Pierce	agarose	isolation of C-terminal peptides

[a]This list is not exhaustive.

Protocol 1. Purification of fibronectin from human plasma on immobilized gelatin

Equipment and reagents

- Sepharose CL-4B (Pharmacia) or suitable alternative
- Cyanogen bromide (Sigma)
- Acetonitrile, AR grade or higher (BDH or alternatives)
- 0.1 M sodium carbonate (Na_2CO_3)
- 0.1 M sodium hydrogen carbonate ($NaHCO_3$), 0.5 M NaCl
- Gelatin, of porcine or bovine skin (type 1) origin (Sigma)
- 0.1 M ethanolamine
- 0.1 M sodium acetate pH 4.5
- Equilibration buffer: either phosphate buffered saline (PBS; 10 mM sodium dihydrogen phosphate, 150 mM NaCl (adjusted to pH 7.5 with NaOH)) containing 0.05% (w/v) sodium azide (NaN_3), or Tris-buffered saline (TBS; 10 mM Tris–HCl, 150 mM NaCl, pH 7.5) containing 0.05% (w/v) NaN_3
- 1 M NaCl
- 8 M urea (AnalaR or equivalent)
- Gelatin–Sepharose (coupled as described in *Protocol 1A* or obtained commercially, e.g. from Pharmacia)
- Citrated human plasma (Sigma or other suitable source) see protocol 5, footnote a.
- Nylon mesh (of suitable pore size to retain gel)
- Chromatographic column (Pharmacia or equivalent) (e.g. 1.2 cm i.d., 12 cm bed height)
- Glass sinter and suction apparatus
- Equilibration buffer containing 1 M NaCl
- 4 M urea in equilibration buffer
- Chromatographic separation system: UV detector (280 nm), peristaltic pump, and fraction collector
- Suitable data logger or chart recorder

A. *Coupling of gelatin to Sepharose CL-4B*

1. Activate 100 ml of Sepharose CL-4B by mixing with 20 g of cyanogen bromide dissolved in 10 ml of acetonitrile at 4°C (3). Remove excess cyanogen bromide by washing rapidly first with cold 0.1 M Na_2CO_3, then with cold 0.1 M $NaHCO_3$ containing 0.5 M NaCl using 10 gel volumes of each solution.
2. Add 400 mg of gelatin dissolved in 100 ml 0.1 M $NaHCO_3$/0.5 M NaCl to the activated gel and gently mix the slurry overnight at room temperature on a mixer.
3. Block unreacted sites, by adding 0.1 M ethanolamine[a], and continue incubation for 1 h.
4. Remove excess gelatin and blocking solution by washing sequentially three times with 10 vols of 0.1 M sodium acetate pH 4.5 and TBS buffers.
5. Wash the gel before use with 10 vols each of 1 M NaCl and 8 M urea. The capacity for fibronectin should be of the order of 1–5 mg/ml gel.
6. Store the affinity media at 4°C in PBS or TBS containing 0.05% (w/v) NaN_3.

B. *Purification of fibronectin*

1. Clean the gelatin–Sepharose media using 8 M urea[b].
2. Equilibrate the cleaned media with several gel volumes of a suitable

buffer such as PBS or TBS until the pH and conductivity are the same as those of the equilibration buffer [c].

3. Filter citrated plasma[d], or an alternative source of fibronectin, through nylon cloth to remove particulates and use immediately.
4. Pack a column with affinity gel at a flow rate of 2 ml/min and then equilibrate it at a flow rate of 1.7 ml/min.
5. Load up to 2 column volumes of plasma on to the affinity gel at a slightly lower flow rate (1.3 ml/min). Alternatively, batch-adsorb the plasma by gently mixing the gel (0.5 vol. of expanded gel:1 vol. plasma) for 1 h at ambient temperature.
6. Wash the gelatin column with 5 vols of the equilibration buffer (at a flow rate of 1.7 ml/min). Loosely bound proteins are removed. Then wash the column with one volume of equilibration buffer containing 1 M NaCl. If using batch-adsorbed gel, wash with 10 plasma volumes of equilibration buffer on a sinter with a fine nylon mesh filter before resuspending in the same buffer and pouring into a suitable column for elution. Higher salt concentrations may also be used for batch adsorption[e].
7. Elute fibronectin using 4 M urea in equilibration buffer[f]. There is no obvious advantage in gradient elution, which produces a very broad peak.
8. Measure the concentration of purified protein by UV absorbance at 280 nm, using an absorbance of 1.28 for a 1 mg/ml solution of fibronectin (1 cm path-length).
9. After cleaning with 8 M urea, re-equilibrate the column in PBS/azide or TBS/azide and store at 4 °C.

[a]Alternatively, glycine or Tris can be used.

[b]Gelatin–Sepharose media should be cleaned using 8 M urea before and after use.

[c]Chromatography may be carried out at room temperature using the buffers indicated. Sodium azide may be added to the various buffers and the source material at 0.05% (w/v) to inhibit microbial growth.

[d]20 mM citrate and/or 50 mM ε-amino caproic acid (EACA) should be added to the PBS or TBS equilibration buffer where there is a risk of clotting and/or plasminogen activation (respectively) in the plasma starting material.

[e]Up to 0.2 M arginine may also be added to the equilibration buffer.

[f]Alternatively, use high concentrations of other chaotropic agents such as sodium iodide, sodium thiocyanate, magnesium chloride, or guanidine–HCl.

Besides chaotropic agents there are a number of less harsh alternatives for elution of fibronectin from immobilized gelatin. The use of 1 M arginine is thought to be the most gentle way of recovering fibronectin (see *Figure 1)* Although this may often be the preferred eluent, it is expensive and contains impurities that absorb at 280 nm, making it difficult to use UV absorbance to monitor protein elution. A urea concentration of 2 M in the presence of 1 M

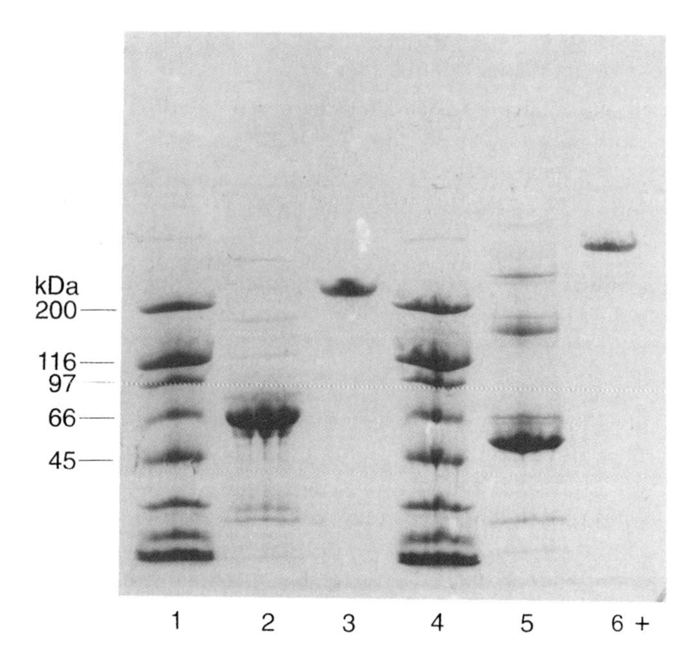

Figure 1. Purification of fibronectin from human plasma using affinity chromatography on gelatin–Sepharose. Gradient (4–15%) SDS–PAGE (Phast system, Pharmacia) was performed under reducing and non-reducing conditions. Lanes 1 and 4, molecular weight markers (200–14.5 kDa, Bio-Rad); lanes 2 and 5, whole plasma, reduced and non-reduced respectively; lanes 3 and 6, arginine-eluted fibronectin, reduced and non-reduced respectively.

sodium chloride (to boost the ionic strength to near that of 1 M arginine), appears to be equally effective. Elution of fibronectin from the immobilized gelatin may also be accomplished by decreasing the buffer pH to below 6, for example by using 0.02 M sodium citrate pH 5.5, or 0.05 M sodium acetate containing 0.1 M NaCl, pH 5.5, or glycine–HCl buffer, pH 2.6. Ethylene glycol, through its ability to disrupt hydrophobic interactions, is also effective for the recovery of fibronectin from gelatin–agarose. Treatment of loaded columns of gelatin–agarose with bacterial collagenase may be used to release fibronectin under relatively mild conditions.

Alternative sources of fibronectin include sub-components of plasma, usually recovered early in the plasma fractionation process, e.g. cryoprecipitate, cold-insoluble precipitate, etc. Some processed plasma intermediates may contain high concentrations of NaCl (up to 3.5 M) which may be directly applied to the gelatin affinity media without difficulty; the reduced non-specific binding to the gel, which is a consequence of the high salt, may reduce the number of wash steps and significantly shorten the purification process. The stability of the source material in high salt should be a consideration in the decision to take this approach.

When used repeatedly, the capacity of gelatin–agarose to bind fibronectin gradually decreases, presumably due to binding of proteins which cannot be eluted. When fibronectin is eluted with urea (3–4 M), the latter can be removed rapidly by direct loading on to an immobilized heparin gel, e.g. heparin–Sepharose, followed by 0.4 M NaCl elution. Alternatively, the urea or arginine eluent can be removed by dialysis at 4 °C against PBS or TBS or by desalting on a suitable gel filtration medium, such as Sephadex G-50 or Fractogel HW55.

2.3 Use of receptor–protein interactions for affinity separations

The ultimate example of the exploitation of natural protein–protein interactions for protein purification has been the isolation of proteins by immobilized protein receptor molecules. This is best illustrated by the use of interleukin-2 receptor (IL-2R) for the isolation of recombinant interleukin-2 (rIL-2) as developed by Bailon and co-workers (4). They have used soluble truncated IL-2R immobilized on a polymeric support to isolate rIL-2 directly from crude fermentor supernatant. The loading buffer is optimized for maximal receptor–ligand interaction and unbound material is removed by washing down in neutral buffer. The pure IL-2 is then eluted with weak acidic eluent which disrupts the receptor–ligand interaction. This procedure is described in *Protocol 2*. The use of immobilized receptor proteins has also been extended to the purification of anti-receptor monoclonal antibodies with therapeutic potential and to the isolation of fusion proteins containing IL-2.

Protocol 2. Receptor affinity chromatography of recombinant interleukin-2 (rIL-2)[a]

Equipment and reagents

- Soluble interleukin-2 receptor p55 subunit (IL-2R)
- Activated affinity matrix (e.g. Nu-Gel AF, Separation Industries) or equivalent
- Chromatography column (G-16 × 25 cm, e.g. Amicon), chromatographic system including peristaltic pump and UV absorbance detector (e.g. Pharmacia)
- Equilibration buffer, e.g. PBS pH 7.4
- Coupling buffer: 0.1 M potassium phosphate, 0.1 M NaCl, pH 7.0
- Blocking buffer: 0.1 M ethanolamine–HCl, pH 7.0
- Elution buffer: 0.2 M acetic acid, 0.2 M NaCl
- Clarified cell culture supernatant containing rIL-2, dialysed into or diluted with PBS pH 7.4
- PBS/azide: PBS containing 0.1% (w/v) NaN_3

A. *Conjugation of IL-2R to activated chromatographic matrix*

1. Wash 20 g of *N*-hydroxysuccinimide (NHS)-activated gel on a glass sinter using deionized water.

Protocol 2. *Continued*

2. Add 28 ml of IL-2R at 2.1 mg/ml in coupling buffer. Incubate for 16 h at 4°C with mixing.
3. Wash the gel twice with PBS and block excess NHS ester groups with 28 ml of blocking buffer, shaking for 1 h.
4. Wash gel in 3 bed volumes of PBS and store in PBS/azide.

B. *Affinity chromatography*

1. Pour the IL-2R affinity matrix into a suitable chromatographic column (19 ml bed 1.6 × 25 cm).
2. Equilibrate the IL-2R column with equilibration buffer at 4 ml/min.
3. Apply rIL-2-containing supernatant at 4 ml/min (160 ml[b]).
4. Wash the column with 5 bed volumes of PBS until the UV absorbance of the eluate returns to baseline value.
5. Elute bound rIL-2 with the eluting buffer (approximately 2 bed volumes)
6. Wash the IL-2R column with 3 vols of PBS and then store it in PBS/azide at 4°C.

[a]Modified from Bailon *et al.* (4) with permission.
[b]A saturating binding capacity is recommended by Bailon *et al.* (4) so as to select only high affinity rIL-2.

The advantage of the receptor affinity chromatographic method over immunoaffinity purification of rIL-2 is that the receptor-purified material appears to be monomeric whereas the rIL-2 purified by immunoaffinity contains substantial aggregated protein.

With the growing number of soluble receptors that are available (both for other cytokines (e.g. interleukin-1, tumour necrosis factor, γ-interferon) and CD antigens (sCD4 etc.), receptor affinity chromatography would seem to have a promising future.

3. Small molecules as ligands for affinity chromatography

Whereas the use of solubilized receptor proteins as affinity ligands is relatively new, the use of immobilized small molecules as purification tools dates back to the beginning of affinity chromatography. Small molecules such as peptides, nucleotides, sugars, and metal ions are important cofactors in the effective activity of a multitude of enzymes, protein hormones, protein carriers, structural proteins, and receptors. These small molecules (or their

Table 2. Some commercially available small ligand affinity chromatographic matrices[a]

Ligand	**Supplier**[b]	**Target for purification**
Heparin	Pharmacia, ACL, TosoHaas, Bioprocessing Ltd, Bio-Rad, Pierce, Sigma, Biosepra	basic proteins, antithrombin, clotting factors
p-Aminobenzamidine	Pharmacia, Pierce, Sigma ACL, TosoHaas, Biosepra	trypsin, other cationic proteases
Lysine	Pharmacia, BioProcessing Ltd, ACL, Pierce, Sigma	plasminogen purification
Glutathione	Pharmacia	glutathone S-transferase (GST) and GST-tagged proteins
Arginine	Pharmacia	prekallikrein; other serine proteases
Biotin	Sigma, Pierce	avidin, streptavidin
N-Acetyl glucosamine	Sigma, Pierce	lectins, glycoproteins
Phospho-gel	Biosepra, Sigma	phosphoproteins, ovalbumin
p-Aminophenyl phosphoryl chloride	Sigma	C-reactive protein
Pepstatin	Sigma	pepsin
Methotrexate	Sigma	dihydrofolate reductase
Polymyxin	Bio-Rad, Sigma	endotoxin
Cholic acid	Sigma	serum albumin
Thyroxine	Sigma	thyroid-binding globulin

[a]This table is not an exhaustive list of possible suppliers or products.
[b]Suppliers' details listed under Sigma refer to entries in the Supelco subsidiary catalogue.

analogues) therefore are ideal ligands for use in affinity-based protein separations (see *Table 2*). Although of broader specificity than protein ligands they are often more robust, more easily and more economically sourced, and more readily lend themselves to generic purification strategies.

3.1 Amino acid ligands

The use of amino acids as immobilized ligands for protein purification has a long history but its value has been vindicated at a preparative scale in the biotechnology industry. At its simplest, the interaction is merely an affinity for a charged or hydrophobic molecule and as such is only broadly specific, hence the term 'pseudobiospecific' coined by Vijayalakshmi (5) for such matrices. A review of manufacturers' catalogues will yield a range of immobilized amino acids for a variety of applications. Other workers have coupled amino acids to generate 'in-house' matrices for novel applications.

Lysine- and arginine-conjugated agarose are widely available and have proven useful in the separation of a number of plasma proteins. For example:

- lysine coupled to agarose via its α-amino group, using cyanogen bromide, can be used in the preparation of plasminogen;

- arginine coupled via a spacer molecule to agarose can be used for the purification of serine proteases and fibronectin (6).

Immobilized tryptophan has been used to purify interferon and cellulases. Vijayalakshmi has promoted the use of immobilized histidine to purify immunoglobulins, although other molecules can also be bound. She attributes its versatility to its ability to undergo charge transfer and its weak hydrophobic nature (7).

Unlike the more sensitive protein ligands, immobilized amino acids can be exposed to harsh cleaning conditions. This is an advantage when working either towards a therapeutic goal or with a very crude feedstock. The use of spacer molecules ensures that steric limitations are kept to a minimum which is especially important considering the small size of the affinity ligands.

An example of the purification of proteins based upon their affinity for amino acids is the purification of plasminogen from human plasma by use of affinity chromatography on lysine–agarose. This method, described in *Protocol 3*, is based upon that of Chibber *et al.* (8). Plasminogen is a 91 kDa, single-chain zymogen found in plasma at a concentration of 60–250 μg/ml. The procedure yields essentially pure plasminogen with only a minor contaminant of albumin which can be easily removed by further processing with anion exchange or dye-affinity chromatography.

Protocol 3. Purification of plasminogen on lysine–Sepharose

Equipment and reagents

- Lysine–agarose either purchased from a commercial supplier (e.g. as Lysine–Sepharose from Pharmacia) or coupled using suitable technology
- Chromatography column: 1.6 cm i.d. × 5 cm, 10 ml (e.g. from Pharmacia or Amicon)
- 0.1 M sodium phosphate buffer, pH 7.4
- 0.3 M sodium phosphate buffer, pH 7.4
- 2 M NaCl
- 0.1 M ε-amino caproic acid (EACA), pH 7.4 (e.g. Sigma)
- Acetone
- Human plasma (Sigma or other appropriate source) see Protocol 5, footnote a.
- Chromatography system including UV monitor, peristaltic pump, and chart recorder or PC interface data logger

Method

1. Pack the column with an aqueous suspension of lysine–agarose resin at the flow rates recommended by the supplier.
2. Equilibrate the column in 0.1 M sodium phosphate pH 7.4 at a flow rate of 20 ml/h.
3. Centrifuge the plasma at 3000*g* for 10 min and filter it through a 0.45 μm pore filter.
4. Apply 6–8 bed volumes of plasma at 20 ml/h.
5. Wash down with 0.3 M sodium phosphate pH 7.4 at the same flow rate until the A_{280} of the eluate falls below 0.05.

6. Elute plasminogen with 1–2 bed volumes of 0.1 M EACA pH 7.4 at the same flow rate.
7. Collect the product and desalt by gel filtration or dialyse to remove the EACA.
8. Clean the column with 2 M NaCl followed by acetone.
9. Re-equilibrate the column in 0.1 M sodium phosphate buffer pH 7.4 or store it in 20% (v/v) ethanol at 4°C.

3.2 Synthetic ligands

The number of synthetic ligands used for affinity purification far exceeds the space available here to describe them. Such ligands are either group-specific agents, which imitate the properties of a true ligand, or they are synthetic adaptations of specific ligands (for instance ligands used for the purification of IgG such as azoarenophiles (1).

3.2.1 Benzamidine

Benzamidine is a synthetic inhibitor of serine proteases and has been used in the purification of trypsin, enterokinase, urokinase, and tissue plasminogen activator (tPA). While early workers synthesized their own benzamidine resins, these are now commercially available from a number of suppliers. With small ligands it is preferable to use a spacer arm between the ligand and the matrix to suppress steric hindrance. Typically the method can be summarized as follows:

(a) equilibrate the resin with a suitable buffer (e.g. 50 mM Tris–HCl pH 8.4);
(b) load the enzyme-containing solution;
(c) wash the column with equilibration buffer until the UV absorbance of the eluate falls to baseline;
(d) elute either with increasing concentrations of sodium chloride or with a specific ligand (e.g. free benzamidine) which can be later removed by dialysis or gel filtration.

HPLC has been applied to separations using benzamidine, for instance in the purification of tPA by Matsuo *et al.* (9). Here they used a silica-based *p*-aminobenzamidine resin (pABA-5PW G5000PW, Toso Haas) which was equilibrated with 50 mM Tris–HCl pH 7.5 containing 0.01% Tween 80. Following the loading of tPA-containing melanoma supernatant, the column was washed and the tPA eluted using 1 M KSCN in equilibration buffer.

More recently, new processing technologies have been applied to separations based upon benzamidine. Powers *et al.* (10) derivatized phospholipids with *p*-aminobenzamidine via a diglycolic acid spacer and produced benzamidine-containing liposomes. These were then used to bind specifically to trypsin prepared in Tris buffer, pH 8.15. The purification was performed on an ultrafiltration membrane and elution was performed using free benzamidine, which could be removed from the released enzyme later by dialysis. Following elution of the bound enzyme, the liposomes can be treated with fresh buffer and stored as a 0.05% suspension at 4°C for further use.

3.2.2 Trifluoromethylketones for the purification of proteases

An example of the use of group-specific ligands was presented by Shiotsuki *et al.* (11) for the affinity purification of serine hydrolases. They described the choice of affinity matrix for the purification of an insect juvenile hormone esterase. A series of analogues of trifluoromethyl-ketones were compared for their relative efficacy, capacity, and ease of use as affinity matrices when immobilized on epoxy-activated Sepharose. If too specific a ligand was chosen then the elution conditions required were too severe and loss of enzyme activity resulted. The generalized purification method is given in *Protocol 4.*

Protocol 4. Use of synthetic ligands for affinity purification of enzymes[a]

Equipment and reagents

- Substrate analogue (e.g. 3-(*S*-substituted)-1,1,1, trifluoro-methylketones (ref. 11))
- Epoxy-activated Sepharose (Pharmacia)
- Ethanol
- 0.1 M sodium phosphate buffer pH 7.4, 5% sucrose, 0.02% (w/v) NaN_3
- Enzyme-containing sample (e.g. juvenile hormone esterase; ref. 11)
- Phosphate buffer containing 0.1% (w/v) *n*-octyl glucoside (Sigma)
- Phosphate buffer containing 0.1% (w/v) *n*-octyl glucoside, 1 mM substrate analogue

Method

1. Couple substrate analogue to epoxy-activated Sepharose according to procedure in ref. 11.
2. Wash the gel sequentially in ethanol, 50% aqueous ethanol, water, then phosphate buffer.
3. Add 50 ml of the enzyme-containing sample to 0.5 ml of drained gel.
4. Incubate with mixing for 8–16 h.
5. Allow the gel to settle and then test the supernatant for residual enzyme activity.
6. Add further enzyme-containing supernatant to the gel and repeat incubation (step 4) until >80% enzyme activity remains in the supernatant.
7. Wash the gel with 50 column volumes of phosphate buffer, then 100 column volumes of phosphate buffer containing 0.1% *n*-octyl glucoside.
8. Elute the enzyme with 20 column volumes of phosphate buffer con-

taining 0.1% *n*-octyl glucoside, 1 mM substrate analogue[b] and mix for 12 h.

9. Repeat until the protein content of the eluate is <5 mg/ml.
10. Concentrate and store the eluted enzyme.

[a]Modified from ref. 11 with permission.
[b]The authors found that salt, pH, or detergent alone failed to elute the bound ligand.

Another example of the use of specific synthetic ligands is the use of calcium-dependent isoquinoline sulfonamide agarose chromatography for the purification of a 36 kDa microfibril-associated protein from bovine aorta (12). Isoquinoline sulfonamide derivatives are calmodulin antagonists and have proven useful for the purification of a number of calcium-binding proteins. The frozen aorta was homogenized in 10 mM Tris–HCl pH 7.5, 2 mM EGTA, 0.5 mM DTT at 4°C. Calcium chloride was added to 2.5 mM and NaCl to 0.2 M. The sample was then applied to a column of isoquinoline sulfonamide-coupled Sepharose (2 cm × 5 cm i.d.) equilibrated in 10 mM Tris–HCl, 0.2 M NaCl, 0.2 mM $CaCl_2$, 0.5 mM DTT, pH 7.5. After a wash with 300 ml of the equilibration buffer to remove any non-specifically bound material, elution was achieved by addition of a calcium chelator comprising 10 mM Tris–HCl pH 7.5, 0.2 M NaCl, 0.5 mM DTT, 2.2 mM EGTA. The recovered protein was judged to be homogeneous on SDS–PAGE (12).

3.3 Nucleotides and cofactors

Some of the earliest ligands to be immobilized for exploitation of ligand–protein affinity were the natural cofactors that are associated with many enzymes and other effector proteins. Examples of such proteins include the NAD-linked enzymes, regulatory proteins (such as protein kinases), and DNA- and RNA-binding proteins. Commercially available nucleotide-conjugated Sepharoses have been available for many years and much work has been published and reviewed previously (13). RNA-binding proteins can be purified using polyuridine (poly(U)) agarose and, since this interaction is mediated via hydrogen bonding, it can be disrupted using a range of chaotropic agents e.g. formamide, resulting in the elution of the purified protein. Raising the operating temperature of the column can also be used as an elution strategy. Immobilized polyadenosine (poly(A)) is marketed as a matrix for the isolation of mRNA-binding proteins, viral RNA, and RNA polymerases. Due to the large size of the ligand considerable capacity is quoted (up to 150 mg/ml of gel) for these resins. Again, elution is via the use of chaotropic agents or by manipulation of ionic strength or pH. Immobilized nucleotide phosphates have been used (for instance adenosine monophosphate and diphosphate where the target protein is an enzyme with a requirement for NAD or NADP) as they can act as structural analogues of the natural cofactor. Bound protein is eluted by manipulation of pH, ionic strength, or temperature, or a low concentration of the competing biospecific ligand may be used (14). The pyrimidine nucleotides have also been used in affinity separations, for example to purify *O*-methyl-transferases such as the sialyl transferase of rat brain. Here the detergent-solubilized membrane proteins were applied to CDP–Sepharose and were eluted by a linear gradient of sodium chloride. Although totally pure enzyme was not obtained initially a second affinity purification on lactosyl ceramide Sepharose yielded homogeneous protein (15).

3.4 Heparin

Heparin is a polydisperse, highly negatively charged mucopolysaccharide with a molecular weight in the range 5–20 kDa and is widely used as an anti-coagulant. Heparin–agarose has been used to separate proteins containing basic surfaces such as hormone receptors, growth factors, and DNA/RNA polymerases (16). In addition to this, heparin–Sepharose has been widely used to separate certain plasma proteins, especially antithrombin, factor IX, and factor XI.

Affinity purification of antithrombin on immobilized heparin was first described in 1974 (17). *Protocol 5*, provided by Lowell Winkelman of Bio Products Laboratory, Elstree, UK, has been used as the basis of a purification scheme for therapeutic grade material (18,19). It illustrates the type of strategy that can be adopted to prepare any of the proteins listed above. Of course, depending upon the target protein and associated contaminants, the choice of equilibration and wash buffers would vary. Elution may be effected by raising the salt concentration or by using a specific competing ligand (e.g. free heparin).

Protocol 5. Affinity purification of antithrombin on heparin–Sepharose

Reagents

- Heparin–Sepharose, either purchased from Pharmacia or coupled using CNBr-activated Sepharose (Pharmacia) and free heparin (type I from Sigma)
- 0.05 M NaCl in 20 mM sodium phosphate buffer, pH 7.3
- 0.15 M NaCl in 20 mM sodium phosphate buffer, pH 7.2
- Human plasma[a]
- 0.40 M NaCl in 20 mM sodium phosphate buffer, pH 7.0
- 2 M NaCl in 20 mM sodium phosphate buffer, pH 6.4
- Chromatographic column (e.g. Amicon) - using a 15 cm bed height (a 4 litre column would have a diameter of 18.4 cm)

Method

1. Adjust the pH of the plasma to 7.2–7.6, if required.
2. Equilibrate/rehydrate the heparin–Sepharose gel with 0.15 M NaCl, 20 mM sodium phosphate buffer.

3. Adsorb plasma or plasma supernatant fraction with heparin–Sepharose gel by gentle mixing at 0–10 °C in the ratio 50:1 (w/v) (i.e. 200 kg plasma is adsorbed on 4 litres of hydrated gel).
4. Pour gel suspension into a chromatographic column and pack the bed.
5. Wash the gel using 4 column volumes of 50 mM NaCl in 0.2 M sodium phosphate then 2.5 vols of 0.4 M NaCl in 20 mM sodium phosphate, both at pH 7.0.
6. Elute the bound antithrombin using 2 M NaCl in 20 mM sodium phosphate (1.5 vols), until baseline absorbance is attained.
7. Clean the gel with 2 bed volumes of acetone followed by 6 bed volumes of deionized water. The gel can be stored with antimicrobial agent added and re-used repeatedly.

[a]Human plasma can be obtained from Sigma or via other suitable sources. Due to the risk of infection only a source which guarantees safety on the basis of viral marker screening should be used, and appropriate laboratory protective procedures should be strictly adhered to.

Chromatography in the conditions described in *Protocol 5* will yield up to 50% of the antithrombin in the starting plasma at a specific activity of approximately 7 i.u./mg protein (>90% pure). By altering the ionic strength of the buffer used to wash the column prior to elution, other proteins can be co-purified. For instance, if the salt concentration of the wash buffer is reduced to 0.35 M NaCl, Factor XI will remain bound to the heparin–Sepharose and will co-elute with antithrombin when the 2 M NaCl elution buffer is added (20).

4. Dye-affinity chromatography

4.1 Background to use of textile dyes

The chance discovery of the co-retention of yeast pyruvate kinase with dextran blue (which contains the dye Cibacron Blue F3GA) when eluting from a Sephadex gel-filtration column (21) opened the way to the study of the interaction of proteins with textile and engineered dye molecules as pseudo-affinity or biomimetic ligands. By far the most well-known interaction is that of the blue dyes (Cibacron and Procion Blue) with both nucleotide-containing proteins and albumin. Anthraquinone dyes are blue, other textile dyes are yellow, orange, or red (containing aromatic azo groups), and green dyes have phthalocyanine moieties.

Dye molecules have many advantages as affinity ligands:

- they are cheap and readily available in bulk quantities;
- they are stable to a range of pH extremes enabling full recovery of activity after cleaning;

- they are easily coupled to matrices via reactive groups and when optimally coupled show low leakage;
- they have a high capacity for a range of proteins.

There are disadvantages to their use, however:

- the specificity is restricted to certain classes of protein so they are not universally applicable;
- the specificity is broad within these groups and so selectivity between bound proteins is not as great as for an immobilized specific ligand;
- dye leakage occurs and may be unacceptable in the desired application.

The purification of proteins using the existing range of commercial textile dyes has been extensively investigated and, where this has proven unsuccessful, engineering of the dye molecules has been undertaken so as to increase specificity and extend the range of application (22). Cibacron Blue shows a specificity for albumin, NAD^+-containing enzymes, decarboxylases, glycolytic enzymes, hydrolases, lysases, nucleases, polymerases, synthetases, and transferases. Cibacron Blue (and Reactive Blue-2) also show affinity for interferon, lipoproteins, α-fetoprotein, and has been used in the purification of plasma protease inhibitors, and many other proteins (23) (see *Table 3*). HPLC applications have also been documented (24). Cibacron Blue F3GA (the dye most frequently used for dye-affinity chromatography) has three structural components: an anthraquinone ring which imparts the colouring, a diaminobenzene sulfonate, and a terminal aminobenzene sulfonate. It has been proposed (25) that the dye molecule mimics the structure of the natural NAD^+ nucleotide cofactor in that anthraquinone acts as an analogue of the adenine, the diaminobenzene of the adenyl ribose, and the triazine ring of the NAD pyrophosphate. However, not all NAD^+-binding proteins bind the dye.

Coupling of the dye to the matrix is via a reactive triazine ring. Dye can be coupled to Sepharose 6B at neutral pH from a 6% NaCl solution. The pH is raised to 10.5 and this permits reaction of the reactive trichlorotriazine

Table 3. Examples of the use of dye-affinity matrices

Dye	Target for purification	Reference
Cibacron Blue F3GA or equivalents	particularly NAD-dependent enzymes; albumin, interferon	26, 28, 31
Procion Red HE-3B or equivalents	particularly NADP-dependent enzymes	26
Procion Brown	tryptophanyl tRNA synthetase; albumin	26
Procion Green	uridine phosphorylase choline acetyltransferase, albumin	26
Remazol Yellow	pre-albumin	27
Cationic dyes	trypsin, kallikrein	32
Procion Navy H3B	alkaline phosphatase	30
Procion Yellow HE-3G	inosine dehydrogenase	26

residue with the ionic form of the hydroxyl groups giving rise to a stable covalent bond. By increasing the level of substitution of dye molecules interactions with more than one dye molecule per protein can be induced, resulting in a higher affinity.

The Procion Red dye has a higher affinity for $NADP^+$-containing proteins than those with the NAD^+ cofactor and so can be used in preference to Cibacron Blue for their purification. Procion Blue and Cibacron Blue F3GA were used to purify lactate dehydrogenase (26), Procion Red to purify formate dehydrogenase, Procion Yellow to purify 3-phosphoglycerate kinase, and Remazol Yellow G6L to purify prealbumin (27) and thyroglobulin.

One of the earliest and most commercially valuable applications of dye-affinity chromatography to protein purification to date is that of the isolation of albumin from human plasma (28); this is given in *Protocol 6*.

Protocol 6. Purification of human albumin by dye-affinity chromatography

Equipment and reagents

- Cibacron-Blue F3GA agarose obtained commercially (e.g. Affi-Blue (Bio-Rad) or Blue Sepharose (Pharmacia)) or coupled via epoxy chemistry (29)
- Chromatographic column (for laboratory-scale separations, 20 cm × 1.6 cm i.d.)
- Low pressure chromatographic system (Pharmacia or equivalent) comprising peristaltic pump and UV detector (280 nm)
- 0.1 M NaOH
- PBS: 150 mM NaCl, 10 mM NaH_2PO_4, adjusted to pH 7.5 with NaOH
- Salt/PBS: 3 M NaCl, 10 mM NaH_2PO_4, adjusted to pH 7.5 with NaOH
- PBS/azide: PBS as above but with 0.05% (w/v) NaN_3
- Human plasma[a], dialysed into 50 mM Tris–HCl pH 7.5

Method

1. Allow adsorbent and buffers to reach ambient temperature.
2. Equilibrate the column with three column volumes of PBS.
3. Load the albumin-containing solution on to the column[b], discarding the breakthrough fraction of eluate.
4. Apply PBS until baseline absorbance at 280 nm is restored (up to 6 bed volumes).
5. Elute albumin using 1–2 column volumes of salt/PBS[c]; collect the eluate until the baseline returns to within 5% of the pre-elution value.
6. Strip column in 0.1 M NaOH (2 column volumes) and re-equilibrate with PBS/azide (2–3 bed volumes) prior to storage at 4 °C.

[a]There is a potential risk of blood-borne viruses in untreated plasma. A cold-ethanol precipitate of plasma (Fraction IV paste) may also be used or other albumin source.
[b]The capacity of the matrix for albumin will need to be measured experimentally but levels of ~20 mg/ml can be expected.
[c]The elution of albumin with NaCl is cost-effective; however, specific ligands (*N*-acetyl tryptophan or octanoate) could also be used for more highly selective elution of albumin.

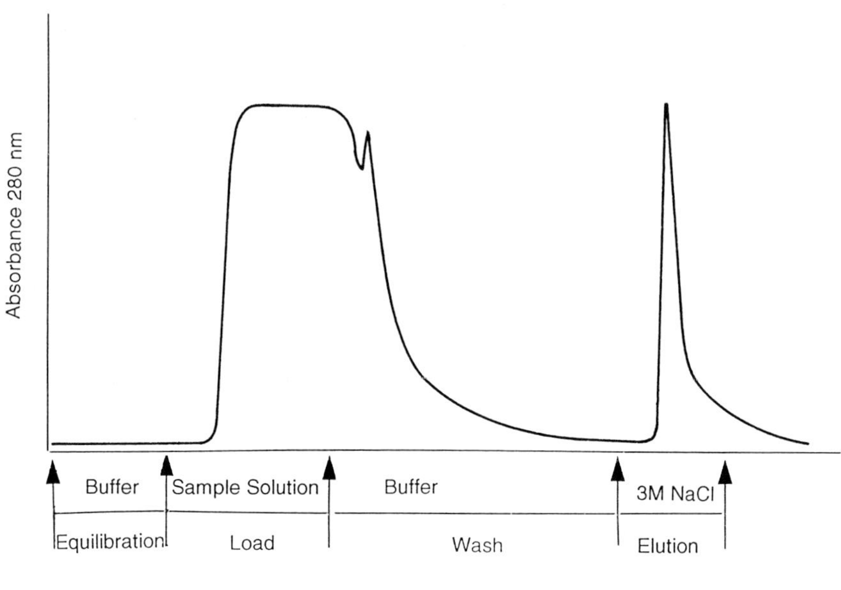

Figure 2. Purification of human serum albumin using Cibacron-Blue dye-affinity chromatography. Chromatographic profile showing equilibration, loading of plasma fraction, wash-down of non-adsorbed material, and elution of purified albumin using 3 M NaCl.

The albumin isolated from plasma using immobilized Cibacron Blue (see *Figure 2*) is >99% pure and the process has been shown to be scalable and capable of yielding material of therapeutic quality (29).

Commercially available test kits of a range of dyes are now available (e.g. PIKSI kit (ACL) or the dye affinity kit from Amicon) so that simple affinity separations can be attempted with a range of different dyes to assess which is most suitable for a given application. These dye-conjugated gels are readily available and are used at industrial scale. Kopperschlager showed that Procion Navy HE-R had a high selectivity and capacity for alkaline phosphatase and phosphofructokinase, using a two-phase aqueous system (30). Birkenmeier also suggested purification strategies for α_2-macroglobulin, albumin, prealbumin, and α_1-acid glycoprotein, based upon immobilized dye-ligands (30a). Another commercially valuable application is the isolation of interferon (31). Cationic dyes have a specificity for trypsin, thrombin, and carboxypeptidases, and it is also possible to separate kallikrein and trypsin on modified cationic dyes (32).

There is a marked range of binding affinities for dyes and, dependent upon the nature of the interaction, either mild or stringent elution conditions are required to recover the protein of interest. For instance, proteins that are less tightly bound via electrostatic and hydrophobic forces can be dislodged by high levels of NaCl, disrupting the electrostatic interaction between dye and protein, Alternatively, chaotropic molecules (such as KSCN, urea, or 10% dioxane) or free natural cofactor (biospecific eluent) may be used. For instance, alcohol dehydrogenase can be bound to Cibacron Blue–agarose from a crude cell extract containing < 0.1 M NaCl in the buffer, then eluted with a gradient of NADH.

4.2 Developments in processing technology

Affinity purification using dye-conjugated liposomes has been described, also affinity ultrafiltration; where using a high molecular weight cut-off membrane, non-specific proteins can be washed out while the specifically bound proteins are retained (33). This has been developed into continuous affinity recycling extraction (34). The dye–protein interaction can be readily adapted for two-phase separations where the distribution of a desired protein in a complex mixture is exploited using a two-phase partitioning between dextran- and polyethylene glycol (PEG)-rich fractions. Usually the PEG component is derivatized with dye molecules as the coupling of dye to PEG proceeds easily under alkaline conditions. By initial use of unsubstituted PEG:dextran the total protein from a dilute feedstock may be concentrated, then with sequential partitioning selective separation of the targeted protein from the proteinaceous contaminants can be achieved with a dye–PEG:dextran system (30a).

The value of expanded-bed dye-affinity chromatography has been demonstrated, in conjunction with use of a novel perfluoropolymeric support, for the purification of dehydrogenases from baker's yeast (36). Expanded-bed chromatography allows use of crude feedstock and rapid flow rates to load the start material and still allows for high performance elution once the bed has been packed. Maltose dehydrogenase could be eluted with a NaCl gradient but glucose-6-phosphate dehydrogenase required elution with the specific free coenzyme $NADP^+$. Scawen and Atkinson have reviewed the large-scale application of dye-affinity chromatography (37).

One problem with existing commercially available dye ligands can be non-specific interactions with the matrix leading to reduced purity. This has been addressed by Galaev *et al.* (38) who chose to coat the matrix with a partially hydrophobic polymer to reduce non-specific interactions while leaving the specific affinity for target protein unaffected. One application of this 'shielded affinity chromatography' uses a thermoreactive polymer so that at elevated temperature the coating binds and prevents non-specific interactions but at a lower temperature the polymer uncoils, resulting in the dissociation of the target protein, elution being achieved merely by a drop in temperature.

Dye-mediated affinity purification can also depend upon the bridging interaction of divalent metal ions, such as zinc. Here, specific retention of target proteins occurs in the presence of the metal ion, and elution results when metal chelators are added (39).

4.3 Strategy for the use of dye-affinity separations

More than one dye-affinity step can be combined in an overall strategy for affinity chromatography. For example Ibrahim-Granet *et al.* (40) purified a metalloproteinase from *Aspergillus* using both a negative and a positive dye separation stage. Due to their relatively broad specificity, dye-affinity matrices lend themselves to use in negative affinity chromatography. Here the protein of interest is not retained by the matrix but the major contaminants are bound, thereby greatly simplifying the later purification steps. Another example of this technique is the removal of albumin from serum samples as a step in the purification of other proteins (e.g. antibodies) which occur at low abundance in plasma or tissuc culture supernatant (see *Protocol 7*, used by kind permission of Dr M. A. J. Godfrey, Clifmar Associates, University of Surrey). When using alternative techniques for monoclonal antibody purification, such as ion exchange or hydrophobic interaction chromatography,albumin is often still present as a contaminant due to its relative abundance in the starting material. By using dye affinity as an initial step, the concentration of albumin is markedly reduced and so does not interfere at later purification steps.

Protocol 7. Negative dye-affinity chromatography on Cibacron Blue-F3GA dye-ligand adsorbent

Equipment and reagents

- Immobilized Cibacron F3GA Blue agarose (Affi-Gel Blue (Bio-Rad) or equivalent)
- Chromatography column (Lab m, no. D823 or equivalent) 5 ml volume, 8 mm diameter
- 0.2 M Na_2HPO_4
- 0.1 M citric acid
- Adsorption and washing buffer: 10 mM Tris–HCl, 15 mM KCl, pH 7.0
- 6 M urea
- 0.1 M NaOH
- Monoclonal antibody containing cell supernatant
- Eluting buffer: 10 mM Tris–HCl, 1.5 M KCl, pH 7.0
- PBS: 20 mM Na_2HPO_4, 137 mM NaCl, 2.7 mM KCl, 1.5 mM KH_2PO_4, pH 7.2
- PBS/azide: PBS containing 0.05% (w/v) NaN_3

Method

1. Allow all the purification buffers to come to room temperature.
2. Clamp column in a vertical position. Cap the outlet and accurately pipette 1 ml of water into the column. Mark the 1 ml level with a waterproof marker on the outside of the column and drain column.
3. Mix adsorbent repeatedly in 5 ml of wash buffer until evenly suspended. Add to column, remove bottom cap, and allow gel beads to pack by gravity, adding more bead suspension gradually until a settled bed volume of 1 ml is obtained (do not allow column to run dry).
4. Add 5 ml of elution buffer to the column and allow to percolate through under gravity.

5. Re-equilibrate column with 5 ml of wash buffer and allow to drain through.
6. Apply antibody-containing solution and allow to percolate through the column under gravity. Collect the breakthrough volume which contains the antibody.
7. Apply a further 0.5 ml of adsorption/wash buffer and combine with the first breakthrough volume.
8. Add 5 vols of elution buffer to the column and discard the albumin-containing eluate.
9. To clean column, add consecutively 5 vols each of 0.1 M citric acid, elution buffer, 6 M urea, then 0.1 M NaOH. Discard all eluate.
10. Add 5 column volumes of adsorption/wash buffer, discarding the eluate. The column is now ready for a further antibody clean-up cycle.
11. Store the column at 4°C after equilibration with 10 column volumes of PBS/azide.

The capacity of the column for albumin will need to be determined empirically although some indication may be provided by the manufacturer. This capacity will govern the load volume of antibody-containing solution on each purification cycle. Removal of the albumin can be confirmed by standard electrophoretic or chromatographic means, prior to the next purification step being commenced.

5. Covalent/thiophilic chromatography

In this section we review the modes of affinity chromatography that exploit the interaction of proteins with sulfur-containing immobilized ligands.

5.1 Covalent chromatography

This technology, based on reversible covalent bond formation, is used to purify those proteins which contain free thiol residues. The first example of covalent chromatography was the use of organomercurial Sephadex by Eldjarn and Jellum to bind thiol-containing proteins. Covalent chromatography was used for methionine-containing proteins by Schechter *et al.* (both cited in ref. 42) who used reactive halogen adsorbent, and eluted the bound protein with β-mercaptoethanol. Two commercial forms of covalent chromatographic resin are sold by Pharmacia: a low-capacity form, Activated Thiol Sepharose 4B, and a high capacity form, 3,2-(pyridyldisulfonyl)-2-hydroxypropyl agarose (better known as Thiopropyl Sepharose). Other suppliers also provide matrices for covalent chromatography. If the protein to be purified has no free thiol groups then these can be induced either by cleavage of disulfide bridges using a reducing agent (β-mercaptoethanol) or by introduction of thiol groups using amino-reactive bifunctional reagents such as SPDP. Binding of protein can occur from Tris or phosphate buffers containing sufficient sodium chloride to prevent non-specific interactions (0.1–0.5 M) and including denaturants such as guanidinium chloride or urea if necessary. The specifically bound protein can then be eluted from the column. The eluents of choice are cysteine, reduced glutathione, or β-mercaptoethanol at 25–50 mM. The resins can also be used to immobilize small molecules containing thiol groups which can then act as specific ligands for the purification of proteins.

Iodoacetyl- and bromoacetyl-derivatized matrices can act as excellent supports for the preparation of thio-containing media (41). (See Chapter 1, Section 7.3). Oscarsson and Porath (42) showed the versatility of the method for isolating plasma proteins by loading serum on to a column of agarose derivatized with pyridine-2-disulfide, equilibrated with 0.1 M Tris–HCl, 0.5 M potassium sulfate, pH 7.5. They found that IgA, IgM, IgG, C3, C4, and α_2-macroglobulin all bound and could be eluted by raising the salt concentration. Other examples include the purification of lecithin cholesterol acyltransferase by Holmquist (43) and of chymopapain by Baines *et al.* (44).

5.2 Thiophilic purification

Thiophilic chromatography was developed by Porath in 1985 (45). This family of ligands (T-gel) results from the treatment of divinyl sulfone-activated agarose with β-mercaptoethanol resulting in the formation of a ligand with one or more thioether residues. The affinity of such residues for immunoglobulins, and more recently for immunoglobulin fragments, has been the basis for a general and effective means of antibody purification (see also Chapter 6). Gels for thiophilic purification are more robust and have specificities equivalent to, or broader than, the predominant immunoglobulin purification affinity techniques which use the bacterial Fc receptor ligands Proteins A and G. Thiophilic resins can bind to antibody fragments that lack the necessary Fc regions needed for Protein A/G binding and so thiophilic purification of antibody single-chain variable region (ScFv) molecules has become popular. These are the smallest antigen-specific subunits that can be made by engineering of the immunoglobulin light and heavy chain genes.

Binding to the immobilized ligand is promoted by high ammonium sulfate concentrations, while desorption is achieved by lowering or removing the salt altogether. However, the interaction is not simply a hydrophobic interaction as can be seen by comparison of the effect of high sodium chloride concentration, which enhances hydrophobic interactions but not those mediated by thiophilic interaction (41). The number of thioether groups incorporated into each ligand molecule may influence the strength and specificity of the interaction. Lihme and Heegaard (46) failed to show affinity for ScFv with a singly substituted T-gel but Schulze (47) demonstrated such affinity with a dithioether-substituted ligand. There are now a number of commercially available matrices but many of the papers describe the synthesis of home-made gels. This method is adapted from Bridonneau and Lederer (48). (with permission from Elsevier Science).

(a) Take Sepharose 6B and resuspend in 0.5 M Na_2CO_3 (pH 10).

(b) Add 0.05 vol. of divinyl sulfone and mix for 3 h at room temperature.

(c) Wash gel with water.

(d) Resuspend in an equal volume of 0.5 M Na_2CO_3 and add 0.1 vols of β-mercaptoethanol and incubate overnight with shaking.

(e) Wash gel repeatedly in water to remove reagents.

(f) Store gel in 0.02% (w/v) NaN_3 at 4 °C.

The purification of a monoclonal antibody from tissue culture supernatant is described in *Protocol 8*.

Protocol 8. Thiophilic purification of monoclonal antibody

Reagents

- Thiophilic gel: Affi-T (Kem-en-Tec) or equivalent, packed as a 5 cm × 1.0 cm i.d. gel in a chromatographic column
- Ammonium sulfate
- Monoclonal antibody (mAb) in serum-free tissue culture supernatant
- 0.05 M Tris–HCl pH 8.9
- 20% (v/v) ethanol

Method

1. Adjust the supernatant by adding ammonium sulfate to a concentration which induces binding but not precipitation of the antibody (determined experimentally but in the range 1–1.5 M) at a neutral to mildly alkaline pH (7.5–9).
2. Equilibrate the Affi-T resin in the same concentration of ammonium sulfate in 0.05 M Tris–HCl pH 8.9.
3. Pass the supernatant through the column at a linear flow rate of 60–120 cm/h. Collect depleted supernatant to ensure that all of the mAb is retained.
4. Elute non-specifically bound proteins with an intermediate concentration of ammonium sulfate in 0.05 M Tris–HCl pH 8.9. The exact concentration of ammonium sulfate needed to remove all contaminant proteins should be determined experimentally[a]. Alternatively, a linear gradient of decreasing ammonium sulfate from 1.2 to 0 M in 0.05 M Tris–HCl may be used.

Protocol 8. *Continued*

5. Elute the mAb using 0.05 M Tris–HCl pH 8.9, without ammonium sulfate.
6. Re-equilibrate column in ammonium sulfate as in step 1 if further separations are to follow or in 20 % (v/v) ethanol if the column is to be stored (4 °C).

[a] The optimum concentration of ammonium sulfate should be used that allows elution of all contaminant proteins while the mAb remains bound.

The advantages of thiophilic purification of antibodies over immunoaffinity or Fc receptor protein-mediated purifications are:

- stable affinity resin which permits robust cleaning methods;
- low cost of thiophilic chromatographic resin;
- broad specificity for immunoglobulins.

Use of Affi-T to purify mAbs is illustrated in *Figure 3* (49).

6. Metal-chelate chromatography

The general principles of metal-chelate chromatography (also known as immobilized metal ion affinity chromatography, IMAC) are described in Chapter 7. An example of this technique is described here, for the purification of coagulation factor IX protein from human plasma.

To appreciate the benefits offered by this method, it is helpful to understand the prior purification difficulties. The concentration of factor IX in plasma is low, both in absolute terms (approximately 4 μg/ml) and relative to the other plasma proteins (approximately 70 000 μg/ml). Furthermore, factor IX belongs to the group of 'vitamin K-dependent' (VKD) glycoproteins which have very similar chemical, physical, and biological properties, due to a high degree of sequence homology and the presence of negatively charged γ-carboxyglutamic acid residues. Although this allows the separation of VKD proteins from other plasma proteins, it complicates the purification of one from another. An additional difficulty is that most VKD proteins are present in plasma as zymogen forms of enzymes which may be activated and/or degraded spontaneously during some purification procedures.

Initially, cryoprecipitation is used to remove fibrinogen which could otherwise form a clot during processing. This is a standard procedure when manipulating plasma (50), yielding fibrinogen-depleted plasma which can then be subjected to anion-exchange chromatography. The ion-exchange step separates VKD proteins from the large amounts of albumin and immunoglobulins which would otherwise compromise the efficiency of subsequent affinity purification. Once the VKD proteins have been isolated from the plasma,

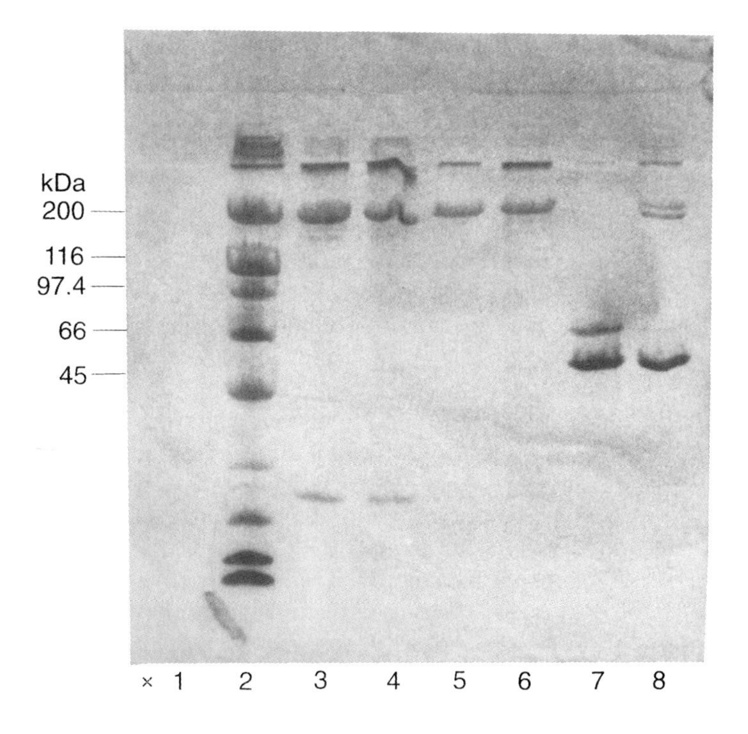

Figure 3. Gradient 10–15% SDS–PAGE (Phast system Pharmacia) of human monoclonal antibody (mAb) purified by thiophilic chromatography.

Lane 1, empty; lane 2, molecular weight markers; lane 3 and 4, purified mAb eluted with 50mM Tris–HCl pH 9.0; lane 5–6, mAb eluted with 0.3 M $(NH_4)_2SO_4$ wash; lane 7, contaminant proteins eluted with 0.7 M $(NH_4)_2SO_4$; lane 8, mAb-containing cell supernatant. From (49) with permission.

they can be purified by affinity chromatography. Metal-chelate chromatography was chosen for this purification in preference to the other options, immunoaffinity or immobilized polysaccharide affinity chromatography, for the reasons described below.

Immunoaffinity purification is monospecific, allowing the purification of only one target protein, whereas advantage may be gained from the purification of several proteins in a single step. The use of antibodies also requires considerable effort in characterization, immobilization, and stabilization to ensure a consistent, re-usable affinity matrix. Recovery of bound protein from the immobilized antibody can require harsh eluents which may damage and/or contaminate the target protein. There is also the possibility of antibody leakage into the protein product, with consequent problems due to inhibition and impurity.

Immobilized sulfated polysaccharides (e.g. heparin or dextran sulfate) can be used to purify more than one VKD protein in a single step. The interaction between matrix and protein is electrostatic and requires adsorption at low ionic strength followed by protein elution at high ionic strength. Given that the VKD-protein mixture is eluted from the primary anion-exchanger at high ionic strength, purification by this method requires an intervening desalting step by dialysis or dilution. After elution from the polysaccharide, a further desalting step is required.

In contrast, metal-chelate chromatography is effective when protein is adsorbed at high ionic strength and eluted at low ionic strength. For the example given, copper ions are immobilized on the chelating gel as this complex has a high capacity for factor IX, yields a concentrated protein eluate and minimizes the risk of metal-ion leakage from the gel. Other metal ions can be used for factor IX purification, though metal-ion leakage is more of a problem. The chelating gel is also stable and can be re-used many times.

The discovery that metal-chelate chromatography could be used to overcome many of these purification problems demonstrates many of the benefits of IMAC alluded to elsewhere in this book. Principles of the present example (*Protocol 9*) using metal-chelate chromatography to purify plasma-derived Factor IX can be applied to materials derived from other natural and recombinant sources. For pharmaceutical applications, the main advantage of metal-chelate chromatography over other affinity systems is that the inorganic chromatography medium is robust.

Protocol 9. Purification of factor IX from plasma

Equipment and reagents

- VKD protein enriched fraction from plasma[a] (see *Protocol 5, footnote a.*)
- Chromatography columns
- Ultraviolet absorbance chromatography monitor
- Buffer A: 50 mM sodium citrate, 50 mM sodium phosphate, 500 mM NaCl pH 6.5 with citric acid
- Buffer B: 50 mM sodium citrate, 50 mM sodium phosphate, 100 mM NaCl pH 7.0 with citric acid
- Buffer C: 50 mM sodium citrate, 50 mM sodium phosphate, 100 mM NaCl pH 4.4 with citric acid
- Buffer D: 10 mM sodium citrate, 10 mM sodium phosphate, 100 mM NaCl pH 7.0 with citric acid
- Buffer E: 10 mM sodium citrate, 10 mM sodium phosphate, 100 mM NaCl, 20 mM glycine pH 7.0 with citric acid.
- Chelating Sepharose Fast Flow (Pharmacia)
- Copper sulfate solution (5 mg/ml, aqueous)

Method

1. Prepare a chromatography column containing Chelating Sepharose in a water medium. Use 2.6 ml of gel per litre of plasma and pack in a column having dimensions which produce a gel bed-height of 12–17 cm.
2. Wash the Chelating Sepharose gel with 5 vols of copper sulfate solution.

3. Wash the Chelating Sepharose gel with 10 vols of Buffer A.
4. Adjust the chloride concentration in the VKD-protein anion-exchange eluate to 500 mM, by adding sodium chloride.
5. Load the anion-exchange eluate on to the copper-charged Chelating Sepharose column. Monitor eluate absorbance at 280 nm.
6. Wash with 7 gel volumes of Buffer A, to remove Factor II.
7. Wash with 6 gel volumes of Buffer B, to remove Factor X.
8. Wash with 7 gel volumes of Buffer C, to remove remaining Factor X, pre-α-trypsin inhibitor, and Protein C.
9. Wash with 5 gel volumes of Buffer D to re-equilibrate to pH 7.0.
10. Elute Factor IX with Buffer E.

[a]Plasma, depleted of cryoprecipitate, is batch-adsorbed on DEAE–Sepharose (Pharmacia), the column is washed with 10 mM sodium citrate, 10 mM sodium phosphate, 100 mM NaCl, pH 6.5, and the VKD proteins are eluted using elevated concentrations of NaCl.

By this purification procedure, factor IX can be purified 10000-fold from plasma, with an overall yield of 27% and a final specific activity of >150 i.u./mg protein (51). The metal-chelate chromatography allows the removal, by extensive washing, of chemical components such as solvent and detergent which may be introduced to inactivate blood-borne viruses. Reduction of blood-borne viruses in plasma products is critical when preparing pharmaceutical products, but should also be considered for the protection of staff working with such materials in the research laboratory.

Metal-chelate chromatography, in the example given, allows the purification of more than one protein from the loading mixture, which contains prothrombin (factor II), factor X, protein C, and factor IX as well as inter-α-trypsin inhibitor in two molecular forms (pre-α-trypsin inhibitor and inter-α-trypsin inhibitor). These can be separated by the various wash buffers (see *Figure 4)* as described in *Protocol 9*. Even in such a complex mixture of proteins, the capacity of the chelating gel is high, providing binding for approximately 200 units (~1 mg) of factor IX per ml of gel. By selective fraction collection, the concentration of factor IX eluted by Buffer E can be >100 units/ml (0.5 mg/ml). After the recovery of factor IX by Buffer E, the high molecular weight species of inter-α-trypsin inhibitor can be removed from the gel by washing with 50 mM EDTA.

The Chelating Sepharose Fast Flow gel is remarkably robust, though it should be packed in the column under pressure, as high concentration/low pH buffers can cause bed distortion if the gel is not fully compressed. Between uses, the gel is washed with 50 mM EDTA pH 7.0 to remove the bound copper and any residual bound protein. Microbiological contamination is minimized by washing and storage of the gel under 70–85% (v/v)

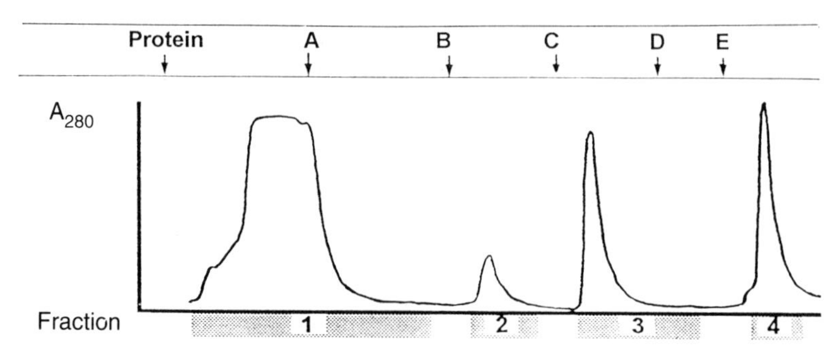

Figure 4. Chromatography of vitamin K-dependent (VKD) proteins using copper-charged Chelating Sepharose. DEAE–Sepharose eluate containing the VKD-protein concentrate was loaded on to copper-charged Chelating Sepharose Fast Flow gel. The gel was washed sequentially with Buffers A–E (defined in the text). The figure shows the elution profile monitored by absorbance at 280 nm. During application of Buffers B and E the sensitivity of the ultraviolet absorbance monitor was increased 10-fold to detect the low concentrations of eluted protein. Fractions contain: 1, factor II; 2, factor X; 3, factor X/Protein C/pre-α-trypsin inhibitor; 4, factor IX.

ethanol at +4°C, which is then removed by copious washing with water before each use. Careful housekeeping of the gel is required to avoid cross-batch contamination by biological components. The presence of highly coloured blue copper is a sensitive visual indicator of the washing efficiency, not usually available when colourless gels are used. This has shown that it is extremely difficult to clean the gel completely when it is packed in the chromatography column. For this reason we routinely unpack the column between uses to ensure effective recycling and sanitization. If handled correctly, the gel may be re-used tens, and possibly hundreds, of times, without loss of performance.

7. Hydrophobic interaction chromatography

7.1 Introduction

Hydrophobic interaction chromatography (HIC) has become such a major topic in its own right that a book could be devoted to this technique alone (52). However, some mention will be made here to this methodology, dependent as it is upon the specific interaction of hydrophobic pockets of amino acids on proteins with non-polar substituted ligands on the immobilized matrix. As with other techniques discussed here this methodology has extended beyond mere simple chromatography to be used in both partitioning and membrane-based formats. At its most basic, an HIC purification may be as described in *Protocol 10*.

Protocol 10. Optimizing hydrophobic interaction chromatography (HIC)

Equipment and reagents

- At least three different HIC matrices with differing hydrophobicities and from a range of major chromatographic suppliers (see *Table 4*). Typically they may be substituted with ethyl, isopropyl, octyl or longer hydrocarbon chains or with aromatic ring structures (e.g. phenyl-agarose).
- Mini columns (<5 ml, of i.d. < 1.6 cm)
- Equilibration buffer (e.g. 0.05 M Tris–HCl pH 7–9) containing either 2–3 M ammonium sulfate or 3 M sodium chloride
- Gel electrophoresis equipment
- Chromatography system (*Protocol 1*)

Method

1. Equilibrate mini columns of each HIC resin in a suitable equilibration buffer.
2. Prepare the protein mixture to be separated so that it conforms to the same composition as the equilibration buffer. Centrifuge off any precipitate that forms. Ensure that the protein of interest is still soluble at the concentration of salt used and, if not, reduce the maximum salt concentration until it is fully soluble.
3. Apply the protein mixture to the selected HIC resins each in turn, continuously monitoring the A_{280}.
4. Wash the columns with 10 column volumes of equilibration buffer. Collect fractions of the unretained material. Test the eluate to ensure that the protein of interest is immobilized.
5. Gradually reduce the salt concentration, preferably over a gradient of 10–20 column volumes until it matches that of the Tris buffer. Collect fractions throughout the gradient and monitor eluted material for the protein of interest.
6. Analyse the protein fractions by electrophoresis (after suitable dilution or dialysis to remove interfering salts) to assess purity.
7. From the initial experiment the HIC resin of choice should be apparent as that to which all of the protein of interest is bound tightly and is eluted in a sharp single peak (largely free of other protein contaminants) on application of a decreasing salt gradient.
8. Optimize the separation by choosing a shallower salt gradient around the range at which the protein of interest elutes and modifying the start condition and sample preparation accordingly. In this way maximal separation of the protein of interest from contaminants can be achieved in a few simple experiments.

Table 4. Some commercially available matrices for hydrophobic interaction chromatography[a]

Ligand	Supplier	Matrix
Methyl	Bio-Rad	polymeric
Ethyl	Sigma	agarose
Phenyl	Pharmacia, Amicon, ACL	agarose, cellulose, synthetic
Propyl	Sigma	agarose
Butyl	Pharmacia, ACL, Amicon	agarose, cellulose
Pentyl	Sigma	agarose
Hexyl	ACL, Sigma	agarose
Octyl	Pharmacia, ACL, Sigma, Amicon	agarose, cellulose
Decyl	ACL, Sigma	agarose
Alkyl	Pharmacia	agarose

aThis table is not intended as an exhaustive listing of potential suppliers or products available.

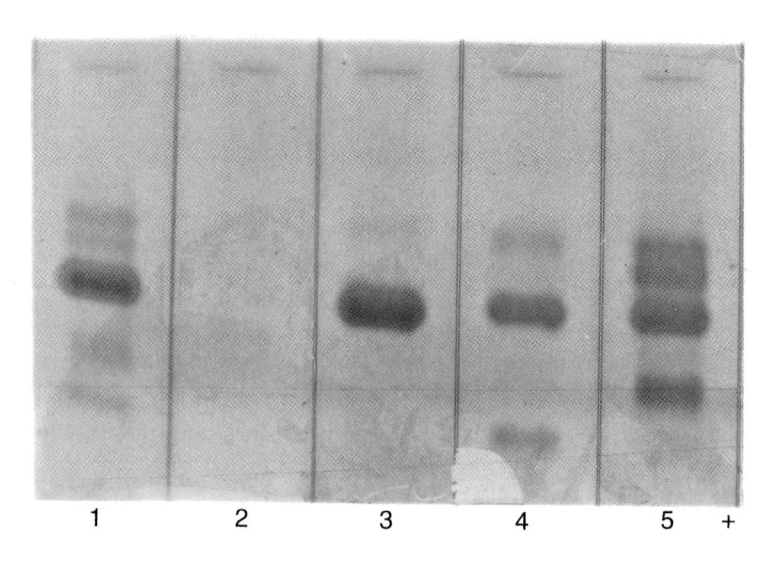

Figure 5. Purification of human α_1-protease inhibitor (α_1-antitrypsin) using hydrophobic interaction chromatography. Purity of column fractions monitored by agarose electrophoresis (pH 8.6 barbitone buffer). Lane 1, starting material: anion-exchange-treated plasma adjusted to 60% saturation with $(NH_4)_2SO_4$; lane 2, 1.2 M $(NH_4)_2SO_4$ wash; lane 3, purified α_1-protease inhibitor (0.9 M $(NH_4)_2SO_4$ eluted peak); lane 4, 0.4 M $(NH_4)_2SO_4$ wash; lane 5 deionized water wash peak. The α_1-protease inhibitor activity was concentrated in the 0.8 M $(NH_4)_2SO_4$-eluted peak, other α-globulins are still present in the other fractions.

7.2 Conventional HIC

This is illustrated by the purification of avian serotransferrin on Phenyl Sepharose by Vieira and Schneider (53). Frozen hen serum was thawed and mixed with 4 vols of 1.5 M $(NH_4)_2SO_4$, 0.5 M sodium citrate pH 6. This solution was loaded on to a 15 ml (1.2 cm i.d.) Phenyl Sepharose column (equilibrated with 1.5 M $(NH_4)_2SO_4$, 0.5 M sodium citrate pH 6). The column was then washed down with 6 bed volumes of equilibration buffer and eluted with a gradient of buffer B (0.75 M $(NH_4)_2SO_4$, 0.25 M sodium citrate, pH 6) (53).

Another example using HIC is given in *Protocol 11* for the purification of α_1-antitrypsin (a highly hydrophilic protein) from an anion-exchange-enriched fraction of plasma (see *Figure 5*). The main contaminants which are present prior to HIC and which are successfully removed by the chromatography are α_2-HS glycoprotein and apolipoprotein A, both more hydrophobic than the α_1-antitrypsin.

Protocol 11. HIC purification of α_1-antitrypsin (AAT) from plasma

Equipment and reagents

- Phenyl Sepharose (Pharmacia) or equivalent
- Chromatographic column (here 18 cm × 4.4 cm i.d.)
- Ion-exchange-purified human plasma[a]
- 2.2 M $(NH_4)_2SO_4$, 20 mM Tris–HCl, pH 7.6
- 0.9 M $(NH_4)_2SO_4$, 20 mM Tris–HCl, pH 7.6
- Refrigerated centrifuge (e.g. Coolspin (MSE) or equivalent)
- 6 M urea (AnalaR or equivalent)
- 20% (v/v) ethanol

Method

1. Take an anion-exchange-purified fraction of plasma (1.1 litre), adjust pH to 7.6 with sodium hydroxide, and add solid ammonium sulfate to a final saturation of 60% w/v (401 g).
2. Stir overnight at 10 °C.
3. Centrifuge in a refrigerated centrifuge (3000*g* at 4 °C), and retain supernatant. Wash precipitate with 200 ml of 60% (w/v) saturated ammonium sulfate and repeat centrifugation. Combine supernatants.
4. Equilibrate Phenyl Sepharose column with 2.2 M $(NH_4)_2SO_4$, 20 mM Tris–HCl, pH 7.6.
5. Apply supernatant to the Phenyl Sepharose column at a linear flow rate of 130 cm/h.
6. Wash column with 2.2 M $(NH_4)_2SO_4$, Tris–HCl and discard both eluates.
7. Elute AAT with 0.9 M $(NH_4)_2SO_4$, Tris–HCl. The ammonium sulfate can then be removed and the AAT concentrated by dialysis and ultrafiltration[b].

Protocol 11. *Continued*

8. Wash off the bound contaminants by elution with deionized water, then 6 M urea.
9. Equilibrate Phenyl Sepharose column in 20% (v/v) ethanol for storage at 4 °C.

[a]Plasma is buffer-exchanged against 35 mM sodium acetate pH 5.2 and then applied to DEAE–Sepharose Fast Flow at high loadings (10 plasma equivalents per ml of gel). AAT is eluted with 0.1 M sodium acetate pH 5.2. See Protocol 5 footnote a.
[b]The success of the purification can be assessed by immunodiffusion and electrophoresis. The purification is incomplete if traces of haptoglobin are still present in the eluted AAT. Losses of AAT also occur into the wash eluates.

Several variations on the basic HIC method have become popular over recent years, including Ca^{2+}-dependent HIC, HIC in alkaline buffers, and preparative HIC. These methods are discussed in the following sections.

7.3 Calcium-dependent HIC

These purifications rely upon the hydrophobic interaction of proteins with an affinity for Ca^{2+}. This hydrophobic interaction is disrupted by the removal of Ca^{2+} using chelators, resulting in the elution of the bound protein. An example is the purification of dog thyroid calcyphosine (54). Calcyphosine is a primary phosphorylated substrate for cAMP-dependent protein kinase following stimulation of dog thyroid cells with thyrotropin. Following labelling of the thyroid cells with [^{35}S]methionine the cells are frozen, homogenized, and separated on DEAE–Sephacel equilibrated with 20 mM sodium acetate, 1 mM DTT, 1 mM EDTA, and protease inhibitors. Following wash-down with equilibration buffer the column is eluted with the same buffer containing 0.2 M NaCl. This step is followed by HPLC anion exchange on DEAE–Memsep (millipore in 50 mM NaCl and eluting with 0.21 M NaCl). This is followed by a HIC step on phenyl TSK (Bio Rad) in 50 mM Tris–HCl pH 7.5 with 1 mM $CaCl_2$. The DEAE-eluted fraction is made 1 mM in $CaCl_2$ and loaded on to the phenyl TSK column. Following wash-down in the same buffer elution of pure calcyphosine is achieved by 1 mM EGTA in 50 mM Tris–HCl pH 7.5. The basis for this purification method is that Ca^{2+} induces exposure of hydrophobic regions on the protein and hence interaction with the HIC resin. The technique was first demonstrated for calmodulin (55)

Batty and Venis (56) demonstrated the purification of calmodulin from Ca^{2+}-dependent protein kinase where the Ca^{2+} dependence was restricted to pH values of <7.5. Hence binding of the target protein to Phenyl Sepharose CL-4B was at pH 7 using 0.1 mM $CaCl_2$, 50 mM Hepes, 1 mM DTT, and 0.5 mM PMSF; elution was achieved with the same buffer containing 1 mM EGTA.

7.4 HIC at alkaline pH

Strong salting-out salts increase the hydrophobic interaction between protein and ligand. However ammonium sulfate cannot be used at alkaline pH; suitable alternative salts are sodium sulfate and monosodium glutamate. Narhi (57) used 2 M monosodium glutamate to purify a mixture of ribonuclease, ovalbumin, and β-lactoglobulin on Phenyl Sepharose. All three were bound under initial conditions at pH 9.5 and were eluted at decreasing concentration of salt.

7.5 Preparative HIC

As well as being a powerful analytical procedure, HIC is capable of being scaled up to purify proteins in quantity. Gagnon *et al.* (58) have reviewed the principles of large-scale HIC. In particular it is a useful method for the separation of endotoxin, viruses, and nucleic acids from proteins produced in a cell culture system. Many of the matrices available at small scale for HPLC separations can be obtained in larger pore size for use at larger scale. The choice of the matrix depends upon the tightness of binding required and the elution conditions appropriate for the recovery of the desired protein at good yield and full activity. HIC is particularly susceptible to changes in buffer composition and the interaction is temperature dependent. It relies on a balance between high salts for binding the protein and having too high a salt concentration which may precipitate the protein irreversibly on the column.

8. High-performance affinity and weak affinity chromatography

The advent of high-performance affinity chromatography (HPAC) in the late 1970s (reviewed in ref. 59) meant that a range of analytical procedures became feasible, finding ready markets with diagnostic and pharmaceutical applications. The quantification of immunoglobulin production by hybridomas using Protein A HPLC resin is a good example (60). Due to the fast cycle time afforded by rigid particulate matrices these columns allowed such methods to compete for turnaround time with the more lengthy immunoassay procedures. Another useful technique disclosed by the advent of HPAC was the use of weak affinity chromatography. Traditionally, affinity separations deal with interactions with an association constant (K_a) of the order of $>10^4$ M^{-1}. However by exploiting weak affinities ($K_a = 10^2$–10^4 M^{-1}) and HPAC it is possible to perform isocratic separations over many column volumes and to obtain separations of molecules which differ only slightly in structure. An example of the use of weak affinity chromatography is the separation of indole and *N*-acetyl tryptophan on immobilized alpha-chymotrypsin (61). Here the difference in affinity of 1.5×10^2 M^{-1} and $1.1 \times$

10^3 M^{-1} can be exploited to achieve a good separation. This technology may have potential in the assay of similar antigens, where these small differences in affinity can be exploited. However, one drawback of weak affinity chromatography is interference due to non-specific adsorption.

9. Stereoselective purification using protein ligands

A recent application of affinity chromatography makes use of a protein's specific interactions with small ligands while immobilized on a matrix, to quantify and separate isomeric molecules. This has been termed reverse affinity chromatography by Mosbach (62). Examples of this have been cited for three proteins in particular: albumin, α_1-acid glycoprotein, and ovomucoid (63). In particular, albumin has affinity for a range of drugs and charged molecules where a stereoselective difference in affinity can be used in the separation or quantification of these agents. Jacobsen and Guiochon (64) describe the use of bovine seum albumin immobilized on ion-exchange resins for the separation of a racemic mixture of tryptophan, mandelic acid, *N*-benzoylalanine, or 2-phenylbutyric acid. One drawback is the gradual leaching of the immobilized protein from the column, which is dependent upon buffer concentration, pH, and any organic modifier present.

10. Conclusions

In this chapter we have visited a number of the techniques that are commonly used for the affinity purification of proteins. While listing specific applications of these methods in the purification of plasma proteins, the general approaches illustrate the principles that underlie these methods. Future developments in affinity separations may be influenced by the development of molecular imprinting (66), where designer matrices can be formed by polymerization of the matrix around the ligand–target complex. Subsequent removal of the target protein yields a highly specific resin for the protein's purification which has the advantages of robustness, cost, and versatility. Diffuse-pore beads and preparative technologies such as expanded beds (see Chapter 1) are already resulting in significant improvements in the process-scale application of affinity separations.

Acknowledgements

We are grateful to Lowell Winkelman (Bio Products Laboratory) and Miguel Godfrey (Clifmar Associates) for permission to use examples from their work.

References

1. Ngo, T. T. (ed.) (1993). *Molecular interactions in bioseparations.* Plenum Press, New York.
2. Ruoslahti, E., Hayman, E. G., Pierschbacher, M., and Engvall, E., (1982). In *Methods in enzymology*, Vol. 82 (ed. L. W. Cunningham and D. W. Frederiksen), p. 803. Academic Press, London.
3. March, S. C., Parikh, I., and Cuatrecasas, P. (1974). *Anal. Biochem.*, **60**, 149.
4. Bailon, P., Weber, D. V., Keeney, R. F., Fredricks, J. E., Smith, C., Familletti, P. C., and Smart, J. E. (1987). *Biotechnology*, **5**, 1195.
5. Vijayalakshmi, M. A. (1989). *Trends Biotechnol.*, **7**, 71.
6. Vuento, M. and Vaheri, A. (1979). *Biochem. J.*, **183**, 331.
7. Vijayalakshmi, M. A. (1993). In *Molecular interactions in bioseparation* (ed. T. T. Ngo), p. 257–75. Plenum Press, New York.
8. Chibber, B. A. K., Deutsch, D. G., and Mertz, E. T. (1974). In *Methods in enzymology* (ed. W. B. Jacoby and M. Wilchek), Vol. 34, p. 425. Academic Press, London.
9. Matsuo, O., Tanbara, Y., Okada, K., Fukao, H., Bando, H.,. and Sakai, T. (1986). *J. Chromatogr.*, **369**, 391.
10. Powers, J. D., Kilpatrick, P. K., and Carbonell, R. G. (1990). *Biotechnol. Bioengng*, **36**, 506.
11. Shiotsuki, T., Huang, T. L., Uematsu, T., Bonning, B. C., Ward, V. K., and Hammock, B. D., (1994). *Protein Expr. Purif.*, **5**, 296.
12. Kobayashi, R., Mizutani, A., and Hidaka, H. (1994). *Biochem. Biophys. Res. Commun.*, **198**, 1262.
13. Potuzak, H. and Dean, P. D. G. (1978). *FEBS Lett.*, **88**, 161.
14. Scopes, R. K. (1989). In *Protein dye interactions: developments and applications* (ed. M. A. Vijayalakshmi and O. Bertrand), p. 97–106. Elsevier, London.
15. Preuss, U., Gu, X., Gu. T., and Yu, R. K (1993). *J. Biol. Chem.*, **268**, 26273.
16. Dean, P. D. G., Johnson, W. S., and Middle, F. A. (eds) (1985) *Affinity chromatography a practical approach.* IRL Press, Oxford, p. 125–131.
17. Miller Andersson, M., Borg, H., and Andersson, L. O. (1974). *Thromb. Res.*, **5**, 439.
18. Smith, J. K., Winkelman, L., Evans, D. R., Haddon, M. E., and Sims, G. (1985). *Vox Sang.*, **48**, 325.
19. Winkelman, L. and Haddon, M. E. (1986). *Thromb. Res.*, **43**, 219.
20. Bolton-Maggs, P. H. B., Wensley, R. T., Kernoff, P. B. A., Kasper, C. K., Winkelman, L., Lane, R. S., and Smith, J. K. (1992). *Thrombosis and Haemostasis*, **67**; 314.
21. Haeckel, R., Hess, B., Lauterborn, W., and Wustar, K. H.(1968). *Hoppe Seyler's Z. Phys. Chem.*, **349**, 699.
22. Lowe, C. R., Burton, S. J., Burton, N. P., Alderton, W. K., Pitts, J. M., and Thomas, J. A. (1992). *Trends Biotechnol.*, **10**, 442.
23. Gianazza, E. and Arnaud, P. (1982). *Biochem. J.*, **201**, 129.
24. Clonis, Y. D. (1989). In *HPLC of macromolecules: a practical approach* (ed. R. W. A. Oliver), p. 157–82. IRL Press, Oxford.
25. Burton, S. J., Stead, C. V., and Lowe, C. R. (1988). *J. Chromatogr.*, **455**, 201.
26. Scawen, M. D. and Atkinson, A. T (1987). In *Reactive dyes in protein and enzyme*

technology (ed. Y. D. Clonis, A. T. Atkinson, C. J. Bruton, and C. R. Lowe), p. 51–85. Stockton Press, New York.

27. Byfield, P. G. H. (1989). In *Protein dye interactions: developments and applications* (ed. M. A. Vijayalakshnmi and O. Bertrand), p. 244. Elsevier, London.
28. Travis, J., Bowen, J., Tewksbury, D., Johnson, D., and Pannell, R (1976). *Biochem. J.*, **157**, 301.
29. More, J. E., Hitchcock, A. G., Price, S., Rott, J., and Harvey, M. J. (1989). In *Protein dye interactions: developments and applications* (ed. M. A. Vijayalakshmi and O. Bertrand), p. 265–74. Elsevier, London.
30. Kopperschlager, G. (1994). In *Methods in enzymology*, Vol. 228 (ed. H. Walter and G. Johansson), p. 121. Academic Press, London.

30a. Birkenmeier, G. (1994). In *Methods in enzymology*, Vol. 228 (ed. H. Walter and G. Johansson), p. 154. Academic Press, London.

31. Knight, E. and Fahey, D. (1981). *J. Biol. Chem.*, **256,** 3609.
32. Clonis, Y. D., Stead, C. V., and Lowe, C. R. (1987). *Biotechnol. Bioengng*, **30**, 621.
33. Mattiasson, B. and Ling, T. G. I. (1986). In *Membrane separations in biotechnology* (ed. C. W. McGregor), p. 99. Marcel Dekker, New York.
34. Pungor, E., Jr, Afeyan, N. B., Gordon, N. F., and Cooney, C. L. (1987). *Bio/Technology*, **5**, 604.
35. Mattiasson, B. (1991). In *NATO Advanced Science Institutes, Series E App. Sci.*, Vol. 204: *Chromatographic and membrane processes in biotechnology* (ed. C. A. Costa and J. S. Cabral), p. 309. Kluwer Press, Dordrecht, The Netherlands.
36. McCreath, G. E., Chase, H. A., Owen, R. A., and Lowe, C. R. (1995). *Biotechnol. Bioengng*, **48**, 341.
37. Scawen, M. D. and Atkinson, T. (1987) In *Reactive dye, in protein and enzyme technology*, (ed. Y. D. Clonis, T. Atkinson, C. J. Bruton, and C. R. Lowe) p. 51–86. Stockton Press, New York.
38. Galaev, I. Y. and Mattiasson, B. (1994). *Bio/Technology*, **12**, 1086.
39. Welch, R. W., Rudolph, F. B., and Papoutsakis, E. T. (1989). *Archiv. Biochem. Biophys.*, **273**, 309.
40. Ibrahim-Granet, O., Bertrand, O., Debeaupuis, J. F., Planchenault, T., Diaquin, M., and Dupont, B. (1994). *Protein Expr. Purif.*, **5**, 285.
41. Hermanson, G. T., Mallia, A. K., and Smith, P. K. (1993). *Immobilised affinity ligand techniques.* Academic Press, London.
42. Oscarsson, S. and Porath, J. (1993). In *Molecular interactions in bioseparations* (ed. T. T. Ngo), p. 403–13. Plenum Press, New York.
43. Holmquist, L. (1987). *J. Biochem. Biophys. Methods*, **14**, 323.
44. Baines, B. S., Brocklehurst, K., Carey, P. R., Jarvis, M., Salih, E., and Storer, A. C. (1986). *Biochem. J.*, **233**, 119.
45. Oscarsson, S. and Porath, J. (1989). *Anal. Biochem.*, **176**, 330.
46. Lihme, A. and Heegard, P. M. H. (1992). *Anal. Biochem.*, **192**, 64.
47. Schulze, R. A., Kontermann, R. E., Quietsch, I., Dubel, S., and Bantz, E. K. F. (1994). *Anal. Biochem.*, **220**, 212.
48. Bridonneau, P. and Lederer, F. (1993). *J. Chromatogr.*, **616,** 197.
49. Ambler, D., Behazid, M., Carr, T., Guest, A., McDonald, D., Lowe, D., Lihme, A., Matejtschuk, P., More, J., Parish, D., Sellick, I., and Stowe, M.(1993). *Chem. Anal.*, June/July, p. 9.

50. Feldman, P. A. and Winkelman, L. (1991). In *Blood separation and plasma fractionation* (ed. J. R. Harris), p. 341. Wiley–Liss, New York.
51. Feldman, P. A., Bradbury, P. I., Williams, J. D., Sims, G. E., McPhee, J. W., Pinnell, M. A., Harris, L., Crombie G. I., and Evans, D. R. (1994). *Blood Coag. Fibrinolys.*, **5**, 939.
52. Builder, S. E. (1993). *Hydrophobic interaction chromatography: principles and methods.* Pharmacia.
53. Vieira, A. V. and Schneider, W. J. (1993). *Protein Expr. Purif.*, **4**, 110.
54. LeCocq, R., Lamy, F., Erneux, C., and Dumont, J. E. (1995). *Biochem. J.*, **306**, 147.
55. Gopalakrishna, R. and Head, J. F. (1985). *FEBS Lett.*, **186**, 246.
56. Battey, N. H. and Venis, M. A. (1988). *Anal. Biochem.*, **170**, 116.
57. Narhi, L. O., Kita, Y., and Arakawa, T. (1989). *Anal. Biochem.*, **182**, 266.
58. Gagnon, P., Grund, E., and Lindback, T. (1995). *Biopharm.*, April, pp. 21, 36.
59. Ohlson, S., Hansson, L., Glad, M., Mosbach, K., and Larsson, P. O. (1989). *Trends Biotechnol.*, **7**, 179.
60. Fulton, S. P., Meys, M., Varady, L., Jause, R., and Afeyan, N. B. (1991). *Biotechniques*, **11**, 226.
61. Ohlson, S. and Zopf, D. (1993). In *Molecular interactions in bioseparations* (ed. T. T. Ngo), p. 15–25. Plenum Press, New York.
62. Birnbaum, S. and Mosbach, K. (1991). *Curr. Opin. Biotechnol.*, **2**, 44.
63. Heard, C. M. and Brain, K. R. (1994). In *Highly selective separations in biotechnology* (ed. G. Street), p. 179. Chapman and Hall, London.
64. Jacobsen, S. C. and Guiochon, G. (1992). *Anal. Chem.*, **64**, 1496.
65. Mosbach, K. and Ramstrom, O. (1996). *Bio/Technology*, **14**, 163.

4

Affinity separations of nucleic acids

RAM P. SINGHAL and CHENNA R. CHAKKA

1. Introduction

Recent advances in affinity chromatography have created a wide variety of important applications for the purification of diverse biomolecules, such as nucleic acids and their components, and antibodies, antigens, and enzymes. The affinity ligand is generally immobilized on an insoluble support and a mixture of compounds containing the molecule of interest is passed over the affinity support, usually in a chromatographic column. Ideally, the molecule of interest, having complementarity to the affinity support, binds to it while other molecules, exhibiting little or no binding affinity, pass through the column unrestricted. Interactions between the affinity molecule and its complement, i.e. the molecule of interest, must be maximized while interactions between the affinity molecule and other molecules in the mixture must be minimized. The next step in the process is to find appropriate conditions for making the complex dissociate, in order to recover the substance of interest without modifying the affinity ligand or the substance of interest. This can be achieved, for example, by increasing the buffer salt concentration, changing the buffer pH, or adding a competing molecule that is structurally similar to the molecule of interest. This article describes methods of affinity chromatography for separating nucleic acids by making use of their interactions with boronate ligands, plant lectins, and synthetic polynucleotides. Interactions between specific proteins and their cognate double-stranded (ds) DNA, and novel methods for their assay by capillary electrophoresis, are also described.

2. Application of boronate-affinity chromatography to nucleic acids

The interaction of boronic acid with vicinal alcohols (*cis*-diols) has been known for several decades. Recently, this reaction has been used to purify important biomolecules (1,2). Applications of phenylboronate matrices include separation of ribonucleosides, ribonucleotides, and ribo-oligonucleotides from their deoxy derivatives in many situations (3,4), as well as separation and assay of modified nucleosides from common nucleosides (5), separation of nucleotides from cAMP (6), assay of benzo(*a*)pyrene–DNA adducts in cells (7), and isolation of nucleotidyl-peptides (8). Moreover, this technique has been used to isolate a specific transfer RNA (tRNA) from a mixture of 19 other tRNA species (1,5,9,10) and to separate capped from uncapped mRNA (11).

The phenylboronate matrix has also been used in clinical analyses. For example, glycosylated haemoglobin (Hb A_{1c}), which is characteristic of prolonged glycaemia, can be resolved with great confidence from other haemoglobin species (12). The high nucleoside content of the intestinal mucosa of gastric cancer patients can be characterized and assayed by boronate chromatography. Hypoxanthine, uridine, and inosine, which are linked to other disorders, can also be characterized by this method. These examples illustrate the powerful application and significance of boronate-affinity chromatography to clinical analysis and basic research.

2.1 Boronate complex formation

2.1.1 The reaction mechanism

Two reaction mechanisms have been proposed for the interaction of boronate with *cis*-diols. In one of these mechanisms, the trigonal, coplanar boronic acid ionizes to form a tetrahedral boronate anion (*Figure 1a*). This anion can then react with a *cis*-diol to yield a cyclic boronate ester (*Figure 1b*; ref. 13). In the other proposed mechanism, a sequential nucleophilic attack of

(a)

(b)

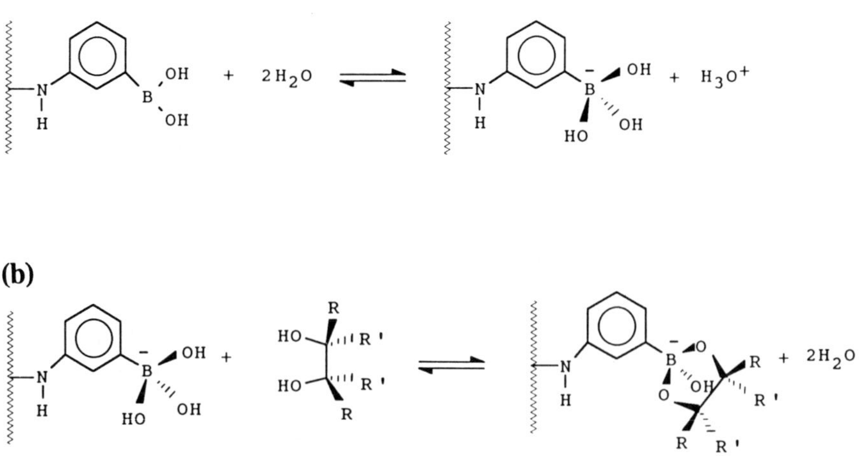

Figure 1. The interaction of phenylboronic acid ligands with 1,2-*cis*-diol compounds. (a) Ionization of the coplanar, trigonal boronate to a tetrahedral structure. (b) Boronate diester formation between a boronate anion and a *cis*-diol (2).

the diol oxygen atoms on the boronic acid causes the formation of an anionic and a neutral species in equilibrium, as shown in *Figure 2* (14). The basic esters with tetrahedrally coordinated boron atoms are found in alkaline solutions. However, only the neutral esters, which hydrolyse in aqueous solutions, are formed at lower pH values. Therefore, the latter type of complex can only be observed in organic solvents.

A reaction mechanism is proposed in *Figure 3* on the basis of ionization and ligand complex formation studied by ^{11}B-NMR and spectroscopic methods (15). Accordingly, the neutral phenylboronic acid (B^0) undergoes facile complex formation in the presence of *cis*-diols, yielding a complexed anion (Bc^-).

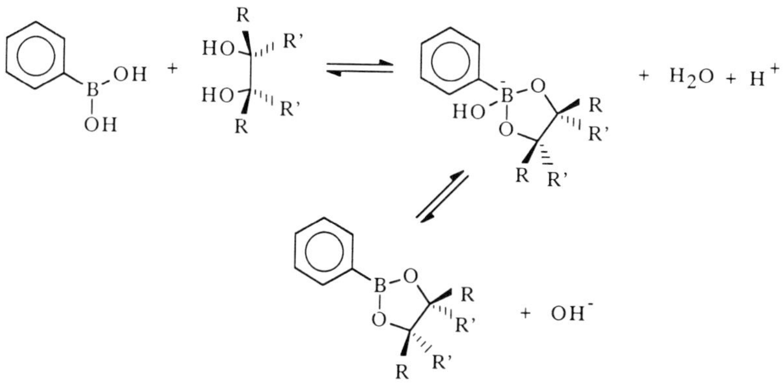

Figure 2. Formation of a complex between neutral benzene boronic acid and *cis*-diols. A sequential nucleophilic attack of the diol oxygen atoms on boronic acid causes the formation of an anionic and a neutral species in equilibrium (2).

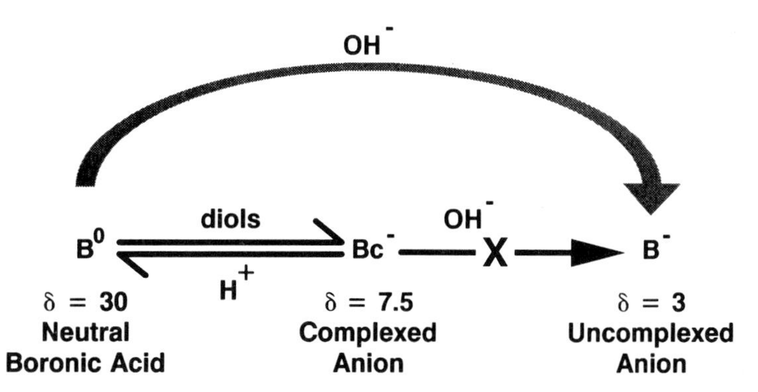

Figure 3. A reaction mechanism proposed on the basis of ionization and ligand complex formation studied by ^{11}B-NMR and spectroscopic methods. Neutral boronic acid (B^0) undergoes facile complex formation in the presence of *cis*-diols, yielding a complexed anion (Bc^-). This complex does not dissociate under alkaline conditions. The neutral boronic acid undergoes ionization under alkaline conditions. (Reproduced from ref. 15 with permission of Elsevier, Amsterdam.)

This complex, formed approximately one pH unit below the ionization of the phenylboronate derivative, does not dissociate under alkaline conditions. Hence, once the complex is formed, both the ligand and the affinity molecule are protected against the adverse alkalinity of the medium. However, the neutral boronic acid ionizes as usual under alkaline conditions (12,15,16).

2.1.2 Reaction conditions to promote complex formation

i. Structural requirements

The interaction of boronates with *cis*-diols is the basis for all separations involving boronates. The major structural requirements for complex formation between boronates and *cis*-diols are that the two hydroxyl groups should be on adjacent carbon atoms and be held in a coplanar configuration. The strength of the interaction increases with the length of the carbon chain in simple polyalcohols in which the hydroxyl groups have the ability to rotate freely around the carbon–carbon bonds. However, the situation is somewhat different in closed-ring carbohydrates, where the configuration of the hydroxyls is stereochemically fixed. Carbohydrates, which normally do not have 1,2-*cis*-diol groups, can still react with borate because of ring flexing and mutarotation. For example, furanoses react more easily than pyranoses because of ring flexing (16,17). The open-chain mutarotation intermediate can form a complex with borate, but the two stereoisomers of glucose (α-D-glucopyranose and β-D-glucopyranose) cannot (18,19).

ii. Reaction environment

Several factors can influence the formation of the ligand-affinity complex. These include pH, nature of the buffer, presence of organic solvents and metal ions, column temperature, and geometry. Higher concentrations of Mg^{2+} and Na^{+} cations appear to stabilize the complex between tRNA and the boronate ligand, but tRNAs can salt out of the solution under these conditions.

iii. Nature of the boronate ligand

A major limitation of the use of phenylboronate ligands for affinity chromatography is the unavailability of an appropriate ligand which can form a complex under neutral or acidic conditions. The most commonly used ligand, 3-aminophenylboronic acid (3aPBA), has a lower ionization constant (pK_a = 8.75) due to the presence of an amine group in the *meta* position. Several attempts have been made to synthesize boronate ligands with lower pK_a values. An electron-withdrawing nitro group has been introduced into the phenyl ring to make boronate more acidic (15,17). Some affinity ligands fail to complex satisfactorily with *cis*-diols in the absence of a spacer arm. In these cases the ligand must be held away from the matrix in order to provide free interactions and avoid steric hindrance. To introduce a small spacer arm and also to offer a functional group for ready coupling with the matrix, the boronate ligand can readily be reacted with an acid anhydride, such as succinic anhydride (15).

2.2 Boronate ligands used for affinity chromatography

A large number of boronate ligands have been synthesized, but only 3aPBA (or its derivative) has been used extensively for immobilization in boronate-affinity chromatography. The amine functionality in this ligand has been very useful for matrix immobilization. A number of coupling chemistries utilize the amino groups of boronate ligands either by introducing a carboxyl group via a spacer arm or by coupling it directly to the matrix. In addition, the 3-amino substitution, as mentioned earlier, reduces the pK_a below that of simple phenylboronate or boric acid. The introduction of an electron-withdrawing group into the phenyl ring stabilizes the tetrahedral boronate anion required for the interaction with *cis*-diols, even at lower pH values.

2.2.1 Determination of binding capacity and dissociation constant by frontal analysis

Frontal analysis is used to determine the binding capacity and apparent binding constant of an affinity matrix. A binding buffer containing the molecule to be captured, such as adenosine at pH 8.5, is applied continuously to a column packed with the affinity matrix until saturation is reached and the effluent concentration of the molecule becomes equal to that of the feed solution. The column is then washed until all the bound molecules are eluted and then a non-binding buffer, such as an acidic pH, containing the molecule and concentration of interest is passed through the column. The elution profile for this kind of analysis reveals important information about the interaction of the ligand and the affinity molecules. The frontal volume (V_f) is given by the difference in the elution volumes between the 50% saturation point for the binding and control experiments.

The binding capacity is calculated from the equation: binding capacity = $V_f[T]/V_t$, where V_f is the frontal volume, V_t is the total column volume, and [T] is the concentration of the feed solution. The apparent dissociation constant can be calculated from the following equation.

$$K_{diss}(\text{apparent}) = V_t[L] - V_f[T]/V_f$$

where [L] is the ligand concentration in the matrix. An example of the frontal analysis of a nitrophenylboronate-affinity column using adenosine is shown in *Figure 4* (15).

2.3 Application of the boronate-affinity method in the study of nucleic acids

2.3.1 Separation of nucleosides and their phosphates

Historically, 5′-nucleoside monophosphates were resolved from their 3′-counterparts by borate complex formation. Among the first molecules to be

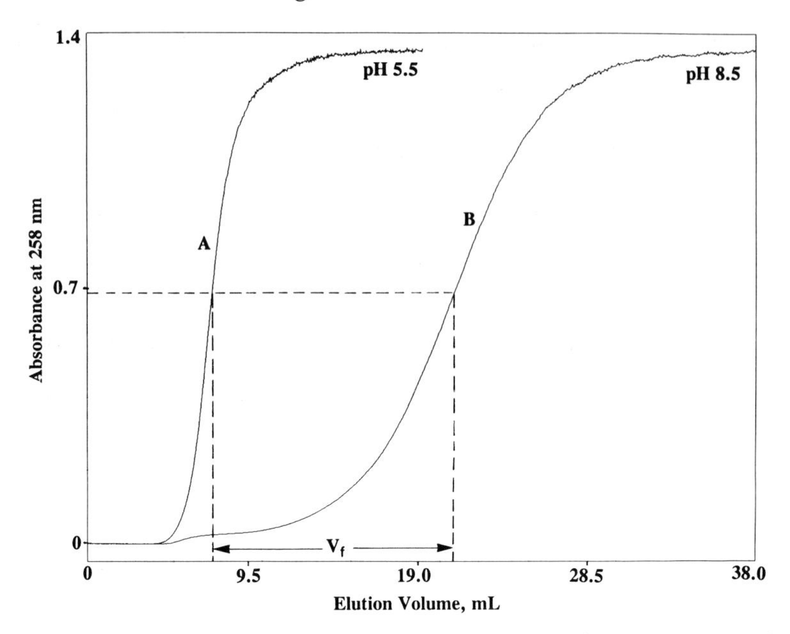

Figure 4. Determination of binding capacity of a boronate-affinity column by frontal uptake of the affinity molecule (adenosine) by the matrix. Curve A was obtained by passing a 0.22 mM solution of adenosine in a 50 mM phosphate buffer, pH 5.5, (a non-complex forming environment). Curve B was similarly obtained using adenosine under complex-forming conditions (pH 8.5). The frontal volume (V_f, 13.7 ml) is derived from the volume difference between the 50% saturation points (mid-points) of the two curves (15). (Reproduced with permission of Elsevier, Amsterdam.)

separated on the immobilized boronate matrices were ribonucleosides and ribonucleotides. The method is also used currently for similar separations (5).

2.3.2 Separation of transfer RNAs

The basis for the separation of aminoacyl-tRNA (aa-tRNA) from non-acylated tRNA is the interaction of the 2′,3′-*cis*-hydroxyl groups of the terminal adenosine of non-acylated tRNAs with the boronate-affinity matrix to yield a stable complex. The binding of uncharged tRNAs is enhanced by using an alkaline buffer. However, alkaline conditions cause hydrolysis of the aa-tRNA ester bond, resulting in uncharged tRNAs which form complexes with the boronate matrix (similar to the other uncharged tRNAs). To achieve satisfactory complex formation at a slightly alkaline or neutral pH, it is recommended that the binding solvent is modified, e.g. by inclusion of high salt content, Mg^{2+}, and ethanol. To prevent non-specific anionic binding to an affinity matrix, such as *N*-[*N*′-(*m*-dihydroxyborylphenyl)succinamyl] aminoethyl cellulose (AEB–cellulose), the matrix should be acetylated after boronate ligation. For example, lysyl-tRNA has been separated from other uncharged tRNAs using *N*-(*m*-dihydroxyborylphenyl)carbamylmethyl-cellulose (CMB–cellulose) and novel reversed-phase boronate (RPB) columns.

Protocol 1. Preparation of boronate-affinity matrices

Equipment and reagents

- Activated charcoal (Norite)
- *m*-Aminophenyl boronic acid
- 6 M NaOH
- 65% (v/v) H_2SO_4
- Polychlorotrifluoroethylene beads, 10 μm diameter (Plaskon CTFE 2300: Allied Colloids; Kel-F: 3M Co.)
- 10 mM sodium acetate, 1 M NaCl, pH 4.5
- 50 mM sodium acetate, 1 M NaCl, pH 4.5
- Dioxane
- Triethylamine
- Decanoyl chloride
- Freeze drier
- Liquid nitrogen 'cold finger' (Aldrich)
- Jacketed glass chromatography column (6.3 mm × 33 mm; model no. LC-6M-13, Laboratory Data Control)

A. *Preparation of the m-aminophenylboronic acid ligand*[a]

1. Dissolve 1 g of *m*-aminophenylboronic acid hemisulfate in 10 ml of degassed water.
2. With mechanical stirring, add 6 M NaOH dropwise to raise the pH to 7.0. Stir the solution for 10 min. A small amount of insoluble precipitate is observed at this stage.
3. Heat the mixture rapidly to 35–40 °C while stirring, then filter immediately. Repeat twice more.
4. Concentrate the filtrate under vacuum to 60% of the original volume and allow to stand at 4 °C overnight.
5. Filter the crystals (Whatman No. 1), wash with *cold* degassed water, and finally dry in a desiccator over 65% sulfuric acid.
6. Redissolve the dried crystals in 10 ml of degassed water, while heating (35–40 °C).
7. Add activated charcoal (0.2–0.3 g) and stir the suspension for 2 min. Filter the solution while hot to remove the charcoal and cool at 4 °C overnight. After two recrystallizations, the product (typical yield: 0.66 g) has a m.p. of 172–174 °C and a λ_{max} at 292 nm at pH 7.0.

B. *Preparation of carboxymethylboronate–cellulose column*

Prepare this matrix as outlined in *Figure 5* using carboxymethyl (CM) cellulose (available from Bio-Rad) as starting material. Approximately 0.3 mmol of boronic acid binds to each gram of matrix. The matrix is acetylated to remove residual amino groups.

The recovery of tRNAs from CMB–cellulose and RPB column matrices is satisfactory (92–97%), but that from AEB–cellulose is incomplete (65–70%). Chromatography on other matrices (benzene boronate linked

Protocol 1. *Continued*

to porous glass, agarose, or Sepharose beads via spacer arms longer than 1 nm) yields little or no separation of aa-tRNA from the uncharged tRNA species, and recovery of the material is very low (about 30%). The low recovery of tRNAs is presumably caused by strong hydrophobic interactions between purine and pyrimidine bases of the polynucleotide and the long spacer arm of the matrix. Such strong interactions are not exhibited by CMB–cellulose and RPB column matrices, which both contain spacer arms of about 1 nm in length (5).

C. Preparation of reversed-phase boronate matrix

1. Treat decanoyl chloride with *m*-aminophenylboronic acid, which is maintained in unprotonated form by carrying out the reaction in dioxane and removing the hydrochloric acid produced by adding triethylamine as shown in *Figure 6*.
2. After completely dissolving 5 mmol of *m*-aminophenylboronic acid in 50 ml of dioxane at 28°C, add 5 mmol each of triethylamine and then decanoyl chloride (3).
3. Stir the solution mechanically for 75 min and then filter to remove any insoluble materials.
4. Concentrate the filtrate by freeze drying and remove traces of dioxane by using a liquid nitrogen 'cold finger'.
5. Suspend the product in 50 mM sodium acetate buffer and then stir at 20°C for 15 h in order to remove any unreacted material.
6. Wash the precipitate with cold water several times to remove salts and dry under vacuum. The yield of the final product is typically 62%. The compound, *m*-(decanoylamino)phenylboronic acid (RPB), exhibits two λ_{max} values at 219 and 246 nm in ethanol, and a m.p. of 121–125°C (5).
7. To coat inert beads with this material, dissolve ~1 g of the ligand completely, after careful drying, in ~20–25 ml of chloroform at 20°C.
8. Mix the solution with 10 g of polychlorotrifluoroethylene beads (10 μm diameter) in a polypropylene bottle and shake in a mechanical shaker for 6 h.
9. Transfer the beads to a glass dish in a fume hood and turn occasionally to remove any chloroform.

D. Preparation of the reversed-phase boronate column

1. Pack the matrix[b] into a jacketed glass column with the help of a spatula and a polyethylene rod of 5 mm diameter.
2. Wash the column with 50 mM sodium acetate buffer, pH 4.5, containing 1 M NaCl to remove any poorly adsorbed affinity groups.

3. Following this treatment, no desorption of the affinity material from the matrix should be observed as monitored at 260 and 280 nm[c].

E. *Application of the reversed-phase boronate column*

RPB matrices have been very useful for separating biomolecules. Both uncharged and aa-tRNAs are retained by a combination of hydrophobic interactions with the ligand (alkyl boronate) and the boronate-affinity functionality (5). To weaken the interactions between aa-tRNAs and the matrix, relatively high concentrations of Cl^- and Mg^{2+} ions are required. However, these ionic conditions are insufficient to break the boronate complex formed between the matrix and the uncharged tRNAs. Strongly held uncharged tRNAs require significantly higher concentrations of Cl^- ions, but only neutral or mildly acidic conditions (a low eluent pH), thus preserving the desired ester bond in aa-tRNA. Uncharged tRNAs bind to the RPB matrix at least one unit below the pK_a value of the phenylboronic acid. The desired aa-tRNA, which fails to complex with the boronate ligand, elutes in the void volume of the column. An example of such separations involving the isolation of liver lysyl-tRNA from other tRNA species is shown in *Figure 7.* The presence of a 1-nm-long spacer arm and the absence of negatively charged vicinal groups in the matrix are apparently responsible for binding at pH 6.8, lower than the usual alkaline pH needed for cellulose- or polyacrylamide-linked phenylboronic acid matrices (5).

[a]Commercially available salt is converted into the free base using a modification of Weith *et al.* (3). See *Protocol 1A* for details.
[b]Reversed-phase boronate matrix is very hydrophobic and hence cannot be suspended in aqueous solvents.
[c]The binding ability of a RPB column, tested for a mixture of adenosine and 2′-*O*-methyladenosine, does not change even after 30 column runs. Since the moiety attached to the matrix is light-sensitive and tends to deteriorate at 20°C, the column should be protected against light and maintained at 4°C at all times.

3. Application of boronate- and lectin-affinity matrices to nucleic acids

3.1 Queuine-containing transfer RNAs

The 'wobble' base, located at position 34, is modified in 61% of eukaryotic and 47% of prokaryotic tRNAs (20,21). Similarly, 31 of 49 mammalian tRNAs contain a modified residue in this position. Guanine (G) is replaced by a modified G residue called queuine (Q), especially at the wobble position, in asparagine, histidine, aspartate, and tyrosine tRNAs. These four specific prokaryotic tRNAs have a simple Q residue which contains a *cis*-diol group, whereas mammalian tRNAs contain either an unmodified Q, for

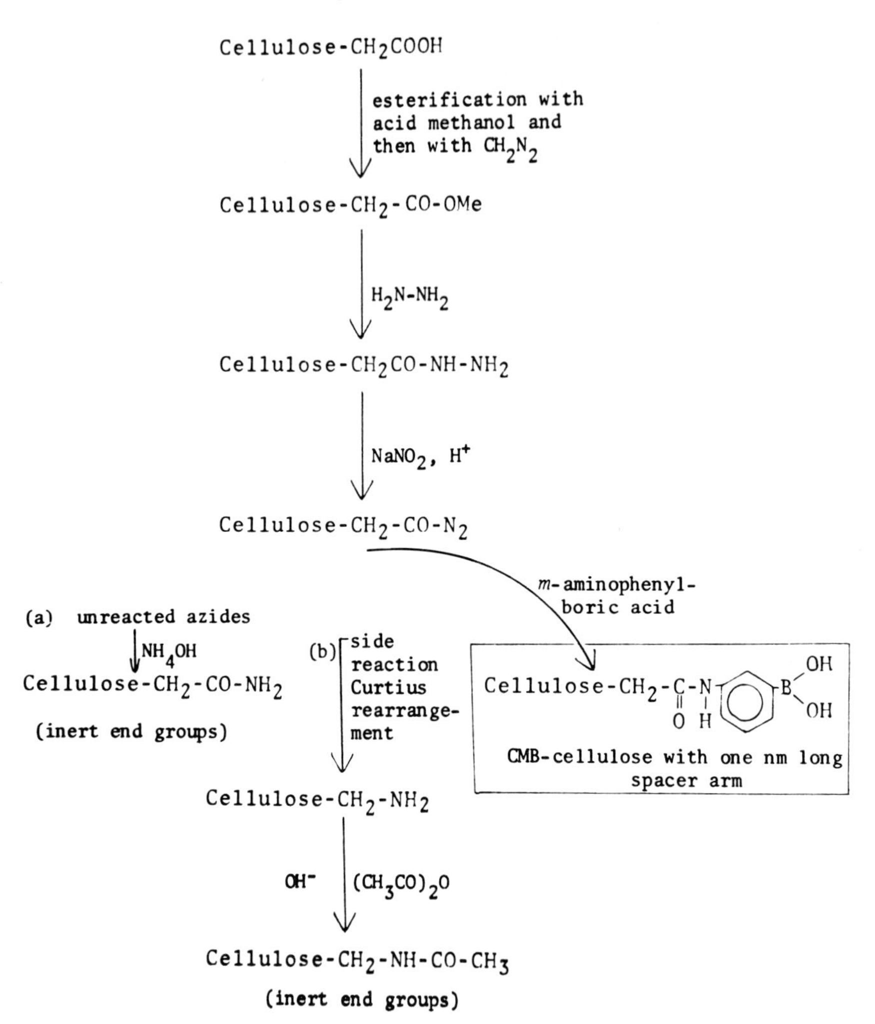

Figure 5. Synthesis of acylated CMB–cellulose matrix. Amine and azide groups produced in side reactions are converted into inert amide groups (reactions a and b). (Reproduced from ref. 4 with permission of Academic Press.)

example in $tRNA^{Asn}$ and $tRNA^{His}$ (*Figure 8*), or a modified Q. In the latter case, a hexose is esterified via one hydroxyl group of the diol in the Q structure, e.g. mannose-Q in $tRNA^{Asp}$ and galactose-Q in $tRNA^{Tyr}$ (*Figure 9*).

3.2 Affinity binding of Q-tRNAs to a boronate matrix

The unsubstituted Q, containing *cis*-diols, reacts with the boronate ligand and forms an anionic complex at a slightly alkaline pH–similar to the reaction of

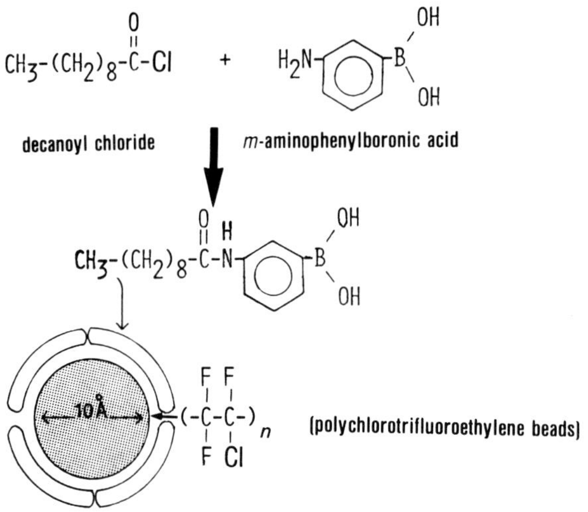

Figure 6. Synthesis of a reversed-phase boronate matrix (5). (Reproduced with permission of Academic Press, Inc.)

ribose *cis*-diols at the 3′-end of tRNAs with a boronate ligand. The bacterial tRNA contains only unsubstituted Q and, therefore, all four Q-containing tRNAs will complex with the boronate matrix. However, only the two mammalian Q-containing tRNAs having unsubstituted Q ($tRNA^{Asn}$ and $tRNA^{His}$) form complexes with the boronate matrix. [The other two mammalian tRNAs with sugar-substituted Q residues are resolved on lectin-affinity matrices specific for hexose (1).] Simple Q-tRNAs tend to complex with the boronate matrix more strongly than those lacking the Q residue in their structure, i.e. complex formation occurs via ribose *cis*-diols. A group separation of Q-tRNAs from other non-Q-tRNA species has been achieved on a CMB–cellulose affinity matrix. This simple separation can be scaled up to process large amounts of tRNAs. For example, Q-tRNAs from mammalian and bacterial sources have been resolved from unfractionated tRNAs on CMB–cellulose columns.

Typically, a 0.5 g sample of bovine liver tRNA, as shown in *Figure 10*, is applied to a CMB column. The column is eluted first with a basic buffer (pH 8.7) and then with an acidic buffer (pH 4.5). Material from each peak is pooled and precipitated with ethanol. The concentrations of different Q-tRNAs in the two peaks are determined by aminoacylation (22). Most simple Q-tRNAs are retained by the CMB column, but only a very small amount of hexose-Q-tRNAs are bound (5). Elution conditions can be examined in order to reduce the interaction between Q-diols and the matrix while maintaining the complex formation between the ribose diols and the matrix.

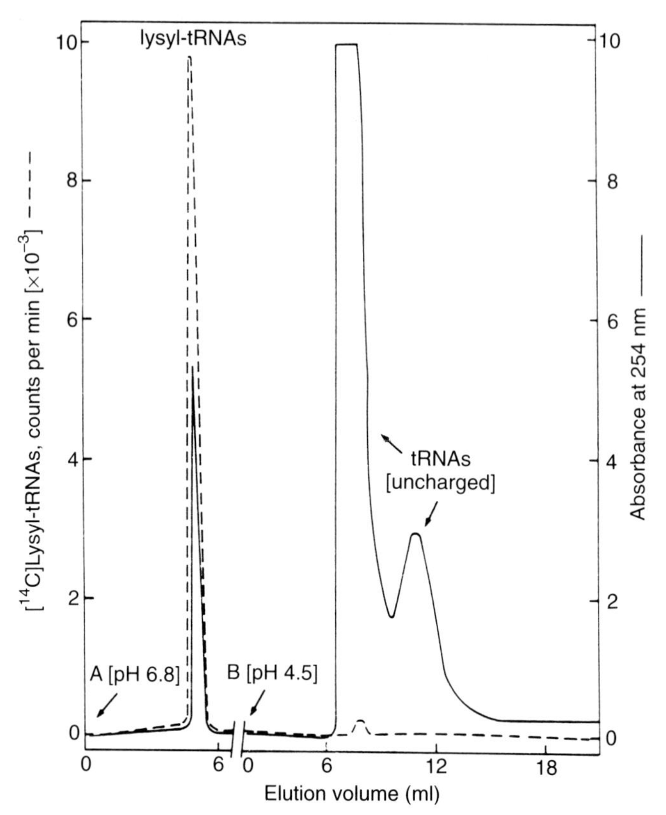

Figure 7. Separation of an aminoacyl-tRNA from uncharged tRNAs by reversed-phase boronate chromatography at neutral pH. About 1 mg of *Escherichia coli* tRNA containing ^{3}H-labelled lysyl-tRNAs (50 000 c.p.m.) in a buffer A of pH 6.8 is applied and then eluted with buffer B (pH 4.5). About 4 nmol of lysyl-tRNA is recovered from other (uncharged) tRNAs. (Reproduced from ref. 1 with permission of Elsevier, Amsterdam.)

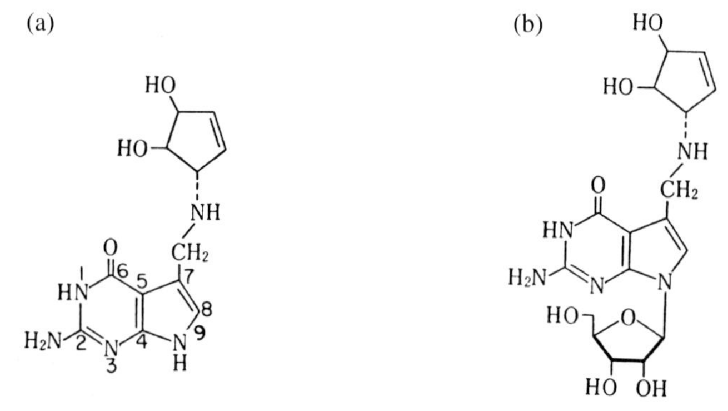

Figure 8. The structure of a modified purine found in the 'wobble position' (position 34) of specific tRNAs. The base, queuine (a), has a dihydroxy cyclopentene ring and tends to form a stable complex with boronate ligands. The nucleoside, queuosine (b), can additionally form a complex via *cis*-diols of the ribose. Unmodified queuosine is found in four specific prokaryotic tRNAs and in eukaryotic asparagine and histidine tRNAs (20).

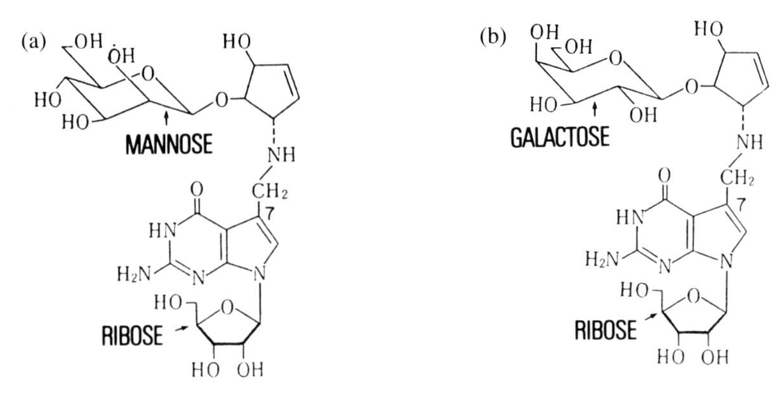

Figure 9. The structure of hexosyl-queuosine of eukaryotic tRNAs. Mannosyl-queuosine (a) is present in liver aspartate tRNA and galactosyl-queuosine (b) in liver tyrosine tRNA. Since cyclopentene *cis*-diols are substituted in hexosyl, they are less effectively retained on a boronate column than those tRNAs containing unmodified Q residues. The hexosyl-Q tRNAs can be easily isolated by lectin-affinity chromatography (1).

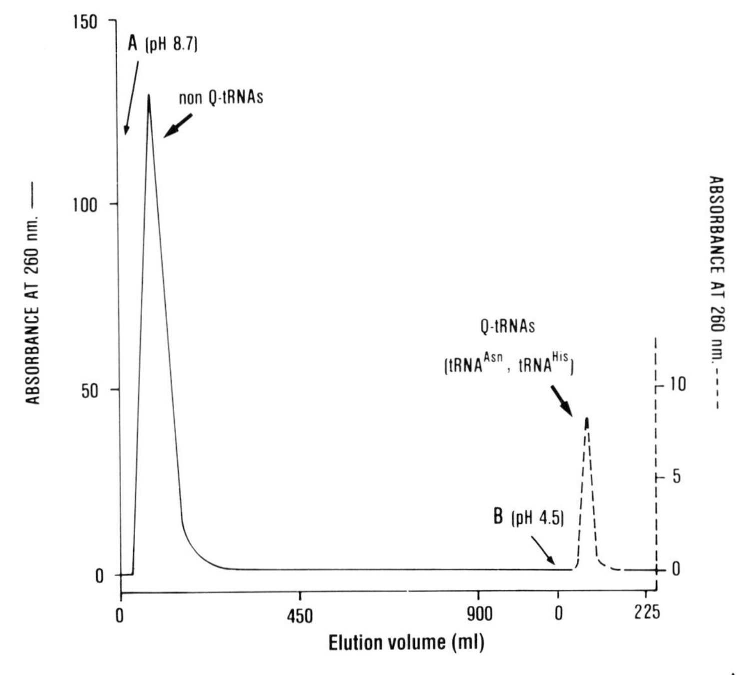

Figure 10. A group separation of unsubstituted-Q-containing tRNAs ($tRNA^{Asn}$ and $tRNA^{His}$) from other tRNA species by boronate-affinity chromatography. Approximately 0.5 g of liver or bacterial tRNAs in 30 ml of buffer A were applied to a CMB–cellulose column (41 cm × 1.5 cm) and eluted first with buffer A (0.7 M NaCl, 0.1 M $MgCl_2$, 50 mM morpholine, pH 8.7) and then with buffer B (0.2 M NaCl, 10 mM $MnCl_2$, 750 mM NaOAc, pH 4.5). The column was eluted at 0.65 ml/min and 10-ml fractions were collected at 4 °C. (Reproduced from ref 1 with permission of Elsevier, Amsterdam.)

Protocol 2. Isolation of $tRNA^{Asn}$ and $tRNA^{His}$ by boronate-affinity chromatography

Equipment and reagents

- Buffer A: 0.7 M NaCl, 0.1 M $MgCl_2$, 50 mM morpholine, pH 8.7
- Buffer B: 0.2 M NaCl, 50 mM CH_3COONa, 10 mM $MgCl_2$, pH 4.5
- CMB–cellulose matrix column: 41 cm × 15 mm, stored at 4 °C (see figure 5)
- tRNA sample (with $tRNA^{Asp}$ removed using Con A–Sepharose (Pharmacia) chromatography)
- Ethanol, cold
- UV detection equipment for monitoring the reaction at 260 nm

Method

1. Equilibrate the CMB–cellulose column at 4 °C with buffer A.
2. Dissolve about 0.5 g of tRNA sample in 30 ml of buffer A (pH 8.7).
3. Apply to the column and elute with the same buffer until all non-bound tRNAs are desorbed.
4. Elute the column with buffer B (pH 4.5) at 0.65 ml/min to release the bound tRNA, collecting fractions of 10 ml and monitoring at 260 nm[a].
5. Recover the Q-tRNAs by reducing the interaction with the acidic buffer B (pH 4.5).
6. Pool these fractions, then precipitate the tRNA with ethanol, and recover it by centrifugation (*Figure* 8).

[a]A moderately basic pH and the high salt content of buffer A reduce both non-specific binding and boronate–ribose *cis*-diol interactions. However, these factors enhance complex formation with boronate-cyclopentene ring diol Q-tRNAs (i.e. $tRNA^{Asn}$ and $tRNA^{His}$).

Protocol 3. Separation of liver $tRNA^{Asp}$ from other tRNAs by lectin-affinity chromatography

Equipment and reagents

- Unfractionated tRNA sample, 0.3 g
- Column (32 cm × 2.5 cm) packed with concanavalin A (Con A)–Sepharose (Pharmacia)
- Buffer A: 50 mM CH_3COONa, 0.15 M NaCl, 5 mM $MgCl_2$, 1 mM $CaCl_2$, and 1 mM $MnCl_2$ pH 6.0[a]
- Buffer B: same as buffer A, but also containing 1 mM 1-*O*-methyl-α-D-glycopyranoside

Method

1. Apply a tRNA sample to a Con A–Sepharose column equilibrated in Buffer A at approximately 0.3–0.5 ml/min. (*Figure 11*). (There is no need for prior removal of other tRNAs using gel filtration.)
2. Elute the column with Buffer A. Under these conditions, all other

tRNAs and RNAs elute in two peaks immediately after the void volume (peaks 1 and 2 in *Figure 11*).

3. Elute the $tRNA^{Asp}$ with Buffer B, which contains competing methyl-α-D-glycopyranoside[b].
4. Clean the column and re-equilibrate with Buffer A.

[a]Divalent cations in the buffer are needed for Con A and sugar (mannose-Q in $tRNA^{Asp}$) interactions.
[b]Aminoacylation of the peak material with appropriate amino acids indicates the presence of only $tRNA^{Asp}$ in the material eluting with Buffer B and the presence of Q-tRNAs ($tRNA^{Tyr}$, $tRNA^{Asn}$, and $tRNA^{His}$) in peak 2. Only very small amounts of these tRNAs are detected in peaks 1 and 3.

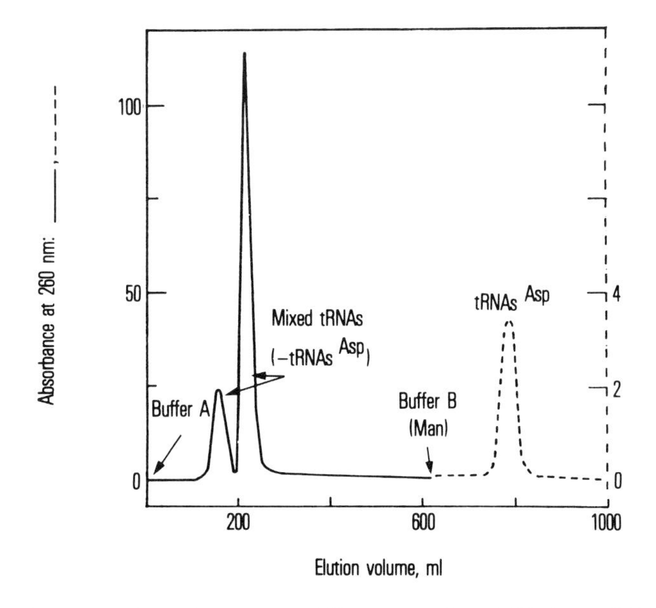

Figure 11. The separation of aspartate tRNA from other tRNA (and RNA) species by lectin-affinity HPLC on a concanavalin A (Con A)–Sepharose column. Note the difference in scale for the two peaks. Approximately 0.3 g of unfractionated tRNA (70%), dissolved in 24 ml of buffer (50 mM NaOAc, 0.25 M NaCl, 10 mM $MgCl_2$, pH 4.5), were applied to a Con A–Sepharose column (32 cm × 2.5 cm; 0.1 mmol Con A per ml of gel). The column was first eluted with buffer A (50 mM NaOAc, 0.15 M NaCl, 5 mM $MgCl_2$, 1 mM $CaCl_2$, and 1 mM $MnCl_2$, pH 6.0). After removing unbound RNAs and tRNAs (peaks 1 and 2), the column was eluted with buffer B (buffer A containing 50 mM mannose or α-methyl-D-glucopyranoside, indicated by an arrow). The column is eluted at 0.5 ml/min, 4°C and 10 ml fractions are collected, pooled, and precipitated by ethanol. (Reproduced from ref 1 with permission of Elsevier, Amsterdam.)

4. Isolation of nucleic acids using polynucleotide-affinity columns

4.1 RNA sample preparation

Phenol extraction followed by ethanol precipitation is the most commonly used method for isolating RNAs. However, alternative methods are now employed for preparing small amounts of RNA. They include guanidine-based procedures and use of sodium dodecyl sulfate (SDS) as a detergent to disrupt cells, and proteases to degrade proteins. Contamination with endogenous nucleases should be avoided during RNA preparation. After RNA isolation, a sample is subjected to agarose gel electrophoresis alongside molecular weight markers to determine the size and quality of the preparation. Oligo(dT)–cellulose or poly(U)–Sepharose chromatography is used to separate poly(A)-containing RNAs (oligomers, probes, or primers) from DNA and other RNA species, including tRNAs and rRNAs. This procedure is adapted from ref. 23, which readers should consult for further details.

4.2 Purification of poly(A)$^+$ RNA

Affinity columns containing oligo(dT)–cellulose or poly(U)–Sepharose are the most commonly used methods of purifying poly(A)$^+$ RNA. Other methods include the use of poly(U) and nitrocellulose membrane filters. A nuclease-free RNA sample is applied to the affinity column under sterile conditions and fractions are collected while monitoring absorbance at 260 nm or radioactivity of the eluate at room temperature. To improve the purity of the product, a second passage of the eluate through the column is desirable. Approximately 10 mg of mRNA is used per gram of column matrix.

Protocol 4. Purification of poly(A)$^+$ RNA on oligo(dT)–cellulose

Equipment and reagents

- Buffer A (oligo(dT) binding buffer): 0.01 M Tris–HCl (pH 7.5), 0.5 M NaCl, 1 mM EDTA, 0.5% SDS. (Also need 2× of buffer A)
- Buffer B (washing buffer): 0.01 M Tris–HCl (pH 7.5), 0.1 M NaCl, 1 mM EDTA
- Buffer C (oligo(dT) elution buffer): 0.01 M Tris–HCl (pH 7.5), 1 mM EDTA
- Buffer D (poly(U) swelling buffer): 0.05 M Tris–HCl (pH 7.5), 1 M NaCl
- Buffer E (poly(U) sample buffer): 0.01 M Tris–HCl (pH 7.5), 1 mM EDTA, 1% SDS
- Buffer F (poly(U) binding buffer): 0.05 M Tris–HCl (pH 7.5), 0.7 M NaCl, 10 mM EDTA, 25% (v/v) formamide
- Buffer G (poly(U) elution buffer): 0.05 M Hepes (pH 7.0), 10 mM EDTA, 90% (v/v) formamide
- Buffer H: 0.05 M Tris–HCl , 2.5 M NaCl, 10 mM EDTA, pH 6.5
- Poly(U)–Sepharose (Pharmacia)
- Oligo(dT)–cellulose (Promega)
- 5 M NaCl or 2× Buffer A
- Ethanol
- 3 M sodium acetate
- 1.5 ml mini columns (New England Biolabs)
- RNA sample of interest

A. *Column chromatography*

1. Suspend oligo(dT)–cellulose (0.1–1.0 g) in 1 to 5 ml of Buffer C, then transfer to 1 to 5 ml mini column, and wash with Buffer A (5 column volumes (CV)).
2. Suspend the RNA sample in Buffer C (1–5 mg/ml), and heat to 65°C for 5 min.
3. Promptly cool RNA sample in ice; dilute with an equal volume of Buffer A (2×). Apply to the column at 0.3–0.5 ml/min.
4. Re-apply the first eluate and then wash the column with Buffer A (5–10 CV).
5. Repeat the wash step with Buffer B (5 CV).
6. Elute the RNA with Buffer C (2–3 CV) and then increase the NaCl concentration in the eluate to 0.5 M using 5 M NaCl solution or Buffer A (2×).
7. Further purify the sample by re-binding it to the column; wash again and re-elute as before.
8. Recover the RNA by precipitation from the final eluate by adding 3 M sodium acetate (0.1 vol.) and ethanol (2.5 vols)[a].
9. Regenerate the column by washing it with 0.1 M NaOH (2–3 CV) followed by distilled water (3–5 CV or until the pH of the effluent is below pH 8), then elute the column with Buffer A containing 0.02% NaN_3 (5 CV).

B. *RNA purification using a batch method*

1. Equilibrate the affinity matrix with Buffer A.
2. Add an aliquot corresponding to 25–300 mg of material to a microcentrifuge tube (1.5 ml).
3. Add the RNA sample, contained in 1 ml of Buffer A, to the tube.
4. After 15 min of gentle shaking, centrifuge the suspension (1500*g*, 5 min) and wash first with Buffer A (1 ml each, 3–5 times) and then with Buffer C (0.4 ml each, 3 times).
5. Pool all eluates, raise to 0.5 M NaCl, rebind, and re-elute three times with Buffer C.
6. Pool all final eluates and recover mRNA by ethanol precipitation.

C. *Purification of RNA on poly(U)–Sepharose*

1. Suspend a poly(U)–Sepharose sample in Buffer D for 1–2 h (20 ml/g; 1 g of poly(U)–Sepharose swells to 4 ml).
2. Wash gel with the same buffer on a sintered glass filter (~100 ml of buffer per gram of Sepharose).
3. Repeat the wash with Buffer G (20 CV) followed by Buffer F (20 CV).

Protocol 4. *Continued*

4. Suspend the RNA sample in Buffer E (1–5 mg/ml).
5. Heat the RNA to 65°C for 5 min, promptly cool in ice, and dilute 5 times with Buffer F.
6. Apply the RNA sample to the column at 0.5 ml/min, then wash with Buffer F (5–10 CV). Because of the high binding capacity of this matrix, the first eluate does not need to be reapplied to the column.
7. Elute the column with Buffer G (1.5–2.0 CV). RNA elutes with the buffer front, causing changes in the gel transparency.[b]

[a]To avoid contamination of the RNA precipitate by SDS, the latter should be omitted from the washing and elution buffers.
[b]The high formamide content of Buffer G interferes with UV absorption measurements of RNA. To maximize RNA yield, effluents are diluted with 0.2 M sodium acetate (4–5×). Presence of formamide may have adverse effects on RNA integrity. This can be avoided by ethanol precipitation (2×) and ether extraction of the aqueous phase (23). The usual purification procedure involves ethanol precipitation of the eluate, but to avoid interference caused by formamide in monitoring the UV, the RNA can be rebound to the column, rewashed and re-eluted with Buffer H, and finally precipitated with ethanol.

4.3 A comparison of oligo(dT)–cellulose and poly(U)–Sepharose matrices

Oligo(dT)–cellulose has the advantage of being relatively rigid and it has a higher binding capacity than poly(U)–Sepharose, thus allowing use of a smaller amount of affinity material. However, because the former matrix contains only fairly short thymidylate chains (18- to 30-mer), it binds mRNA less efficiently, cannot separate mRNAs with different poly(A) chain lengths, and may result in impure samples. Poly(U)–Sepharose, on the other hand, contains long (100-mer) polyuridylate chains allowing enhanced binding and fractionation of different mRNAs. However, the matrix is not rigid and exhibits lower affinity per unit mass, requiring larger columns and making batch procedures difficult.

5. Protein–DNA binding

The interactions between specific proteins and dsDNA are commonly examined to study the control of transcription and gene expression. The interaction between a specific protein and the dsDNA sequence to which it binds is highly selective; factors involved in this interaction are summarized below:

- Hydrogen bonding or van der Waals interactions occur between the polypeptide chain and exposed edges of base pairs, primarily in the major grove of B-DNA.
- Local conformation and configuration of the macromolecules contribute to this interaction—an indirect readout in the recognition process.

- Sequence-dependent binding preferences of the DNA limit the energetically favourable conformations of a particular binding site, therefore imposing additional sequence-dependent constraints on the binding affinity.
- Direct binding energy (6–15 bp) is insufficient to form a stable complex. Additional energy may be provided by direct electrostatic interactions between basic amino acids and the negatively charged sugar–phosphate backbone.
- Sequence-selective binding affinity results from the difference in the binding energies between sequence-dependent and sequence-independent binding.

These interactions are generally examined by gel electrophoresis. Protein-bound dsDNA moves more slowly than corresponding unbound DNA in an electrical field. Though this 'gel mobility shift assay' is commonly used, it is not quantitative and suffers from interference in the form of non-specific binding of dsDNA to other proteins. A typical separation of a dsDNA probe complexed with nuclear proteins, both in the presence and absence of non-specifically binding DNA, is shown in *Figure 12*. In each lane, presence of a complex between the specific DNA probe and the Sp1 protein is evident. However, the presence of DNA complexes with other proteins is also noted. Capillary zone electrophoresis can be used to measure specific protein–DNA interactions (see below) (24).

6. Applications of capillary electrophoresis

6.1 Introduction

Capillary electrophoresis (CE) is a powerful new analytical separation technique that brings speed, quantification, reproducibility, and automation to the inherently high-resolution but labour-intensive methods of electrophoresis (25). Electrophoresis is the separation of charged molecules based on their movement through a fluid (usually a buffer solution called the 'carrier' or 'background electrolyte') under the influence of an applied electric field. Generally the separation is performed in a medium such as a semi-solid slab gel. This separation technique is essentially based on differences in the net charges and shapes of the components to be analysed. Unlike conventional electrophoresis, the capillary technique does not require a supporting medium and the analysis can be carried out in aqueous solutions. Some of the particularly important attributes of CE separations include the large surface-to-volume ratio and low viscosity of the supporting electrolyte. Capillary electrophoresis can be up to twice as efficient as high-performance liquid chromatography. Although electrophoretic techniques vary, all are based on one of the following four electrophoretic modes: moving-boundary electrophoresis (MBE), zone electrophoresis (ZE), isotachophoresis (ITP), and isoelectric focusing (IEF).

In conventional capillary zone electrophoresis (CZE) two competing

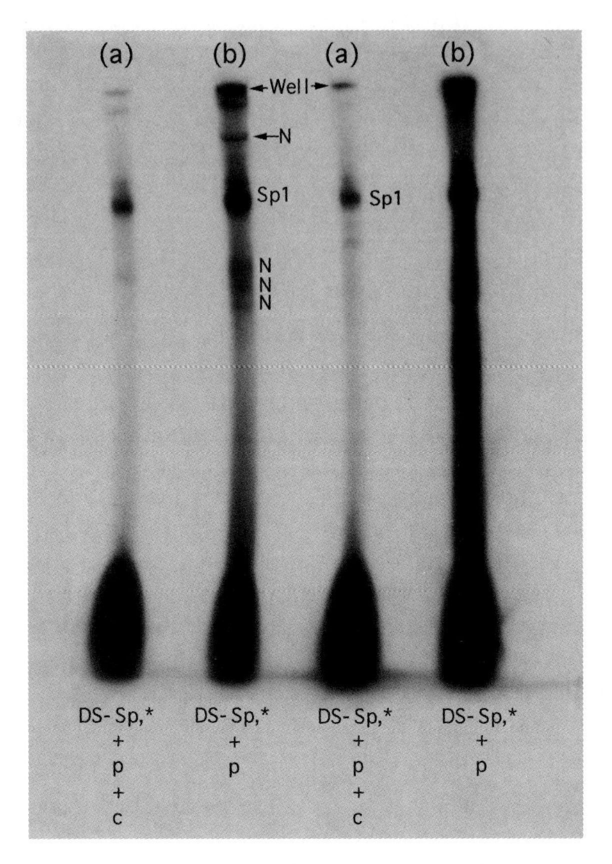

Figure 12. (a) A mixture of dsDNA (15 ng, ^{32}P-labelled, 2×10^5 c.p.m.), non-specific competitor, poly(dI:dC) (2 mg), and nuclear proteins (20–25 mg, nuclear extracts from mouse fibre cells). Note specific binding between dsDNA and Sp1 nuclear protein. (b) A mixture of dsDNA and nuclear proteins. Note non-specific binding with other proteins in the absence of competing DNA. Polyacrylamide gel electrophoresis is performed in a low salt, 9% native polyacrylamide gel and 45 mM Tris–borate–EDTA buffer for 3 h. Abbreviations: DS-Sp_1*, ^{32}P-labelled dsDNA–protein complex, P, nuclear proteins. C, non-specific competitor, poly(dI:dC) added.

forces are involved. While electro-osmotic flow moves analytes towards the cathodic detector, the overall mobility depends on the nature of the analyte's charge. The latter mobility can either add to the migration, if the molecule behaves as a cation, or it can reduce the effective migration if the molecule behaves as an anion. Since nucleotides contain negative phosphate groups, their effective migration in CZE is reduced.

Micellar electrokinetic capillary chromatography (MEKC), a modification of CZE, involves the addition of a surfactant such as SDS in order to yield micelles with a hydrophobic interior and a charged exterior. Neutral and also charged molecules, such as bases and nucleotides, are resolved by MEKC because of differences in their partition coefficients between the bulk of the solution and the micelles.

Capillary gel electrophoresis (CGE), a combination of slab-gel electrophoresis and CE, consists of a solid gel network covalently anchored to the capillary surface. The environment offered to polynucleotides by the gel matrix is similar to that observed in slab-gel electrophoresis. The separation of polynucleotides in CGE is based on a combination of size exclusion and electrophoretic movement towards the anodic detector. However, the electrophoretic migration contributes very little to the effective mobility of the molecule. For example, smaller oligomers carrying fewer negative charges emerge before larger ones.

The separation of purine and pyrimidine bases and also that of nucleosides by CE has been rather limited because these molecules bear no charge. However, these molecules have been successfully resolved by MEKC because they differ in hydrophobicity (26–28). Nucleoside phosphates, carrying one or more negative charges, have been resolved by CZE and MEKC and with certain polymers in the separation medium. Though smaller oligonucleotides have been resolved by MEKC, larger oligonucleotides, polynucleotides, and DNA fragments can be resolved only by CGE and solution polymer capillary electrophoresis (27,28).

6.2 Factors affecting capillary electrophoresis

The resolution, sensitivity, and selectivity of CE for separating oligonucleotides depends on several practical parameters. For quantification, peak area should be corrected for peak velocity. The product of peak area and peak velocity (not just the peak area) should be employed as the quantitative response with the units of area per unit time. The detector response to a high concentration of oligonucleotides varies linearly with increasing sample concentration, but the response to a lower concentration is erratic. An internal standard should be used to avoid this problem. Though sample conductivity (high salt content) significantly decreases detection sensitivity, pH change has only minor effects. The sensitivity increases in excess of one order of magnitude if the sample contents are increased significantly, causing only an insignificant loss in resolution. An increase in the column temperature directly decreases the solute migration (by 1.1% per °C).

6.3 DNA–protein binding studied by CE

6.3.1 Type of detector

Radiolabelled DNA fragments, commonly used in the mobility shift assay technique, cannot be utilized for the CE method. Though an on-line radioisotope detector is available for CE, it cannot be used effectively for the analysis of ^{32}P-labelled DNA (24). The detection limit of the β-ray detector strongly depends on the experimental conditions of the sample employed (quenching) and the CE parameters selected. For CE separations which involve use of a constant high voltage, the detection limit is about 5 nCi, corresponding to about 0.84 fmol of [β-^{32}P]ATP or just a few counts per minute. For the detection of DNA–protein complexes, however, the radioactivity of the sample solution must be at least 15 000 c.p.m./ml, because Sp1 protein exists in a low concentration in the nuclear extract. The loss in sensitivity is caused by the narrow bore of the capillary column (75 μm).

A different approach is employed for detecting DNA–protein complexes in CE. This involves the use of dsDNA labelled with an appropriate fluorescent tag. The binding parameters are measured and the complex quantified by a laser-enhanced fluorescence emission method (29). This method of detection has a limit of detection of about 10^{-21} mol.

6.3.2 Sensitive measurement of DNA–protein complex formation

A fluorescently labelled dsDNA probe is added to a mixture of nuclear proteins, and forms specific DNA–protein complexes. An example of the interaction of the human cardiac actin promoter binding protein with a dsDNA probe is shown here. The separation in *Figure 13* was achieved by the MEKC mode of CE and using laser-induced fluorescence detection (30). Panel 1 shows the migration of a single-stranded DNA (ssDNA, 1.13 nmol/ml), which contains the binding site for serum response factor (SRF, i.e. human cardiac actin promoter). It is labelled with fluorescein at the 5′-end of the sense strand of the probe: [5′-Fl]-GGGGA**CCAAATAACC**CAAGC. Panel 2 shows the migration of dsDNA (1.39 mg/ml), i.e. the sense strand, complexed with an antisense strand, 5′-GCTTG**CCTTATTTGG**TCCCC. (Specific recognition site in bold face). Panel 3 shows the separation of all three components, i.e. ssDNA, dsDNA probe, and the DNA probe (0.17 ng) complexed with a specific protein from a nuclear extract from the mouse myoglobin cell. The CE separations were carried out in a 25 mM $Na_2B_4O_7/H_3BO_3$ buffer, pH 9.0, containing 20 mM SDS, using a hydrodynamic injection technique (for 10 sec) on a 15 cm capillary and with a potential of 10 kV. The detector consisted of an argon laser (488 nm) and a photomultiplier tube operating at 97%.

7. Conclusion

The boronate-affinity chromatography method offers promise as a rapid, easy purification method for numerous biomolecules. However, the current technique suffers from lack of satisfactory boronate ligands and ready-made stable column matrices. Only one boronate derivative, 3-aminophenylboronic acid (3aPBA), is commercially available for immobilization. Different commercial matrices contain a derivative of 3aPBA (or simply a boric acid gel). 3aPBA, which is unstable under alkaline conditions, can, paradoxically, only function satisfactorily under alkaline conditions for a large number of

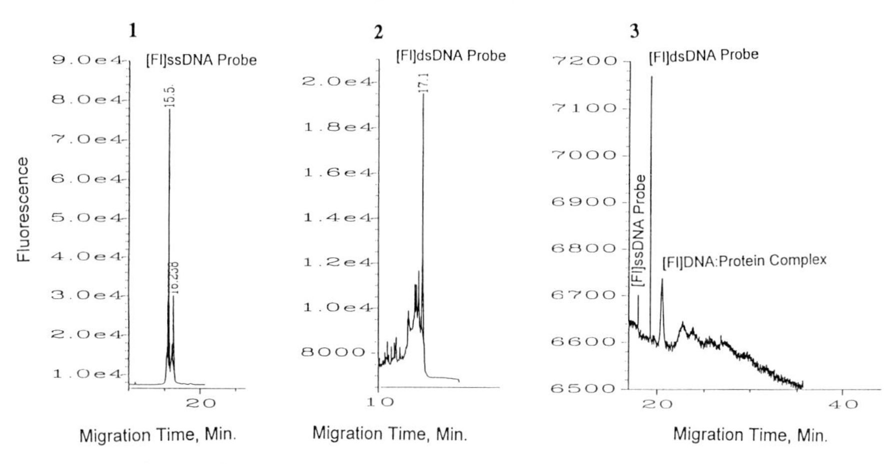

Figure 13. CE-MEKC separation of DNA and DNA–protein complex. CE profiles of (1) fluorescein-labelled ssDNA probe, (2) fluorescein-labelled dsDNA probe and (3) fluorescein-labelled dsDNA probe–protein complex are shown.

biomolecules. Moreover, many biomolecules are also unstable under the alkaline conditions required in chromatography. Furthermore, matrix immobilization techniques also often require alkaline reaction conditions. To overcome the difficulties associated with 3aPBA, novel ligands and coupling chemistries have been described (12). The availability of new, carefully designed boronate ligands can yield highly desirable affinity matrices for the isolation of a great number of biological substances whose purification in molecular biology as well as in clinical chemistry is in increasing demand.

Oligo(dT)–cellulose or poly(U)–Sepharose chromatography is used to purify poly(A)$^+$ RNAs. The merits and disadvantages of these two matrices are discussed above.

The interactions between a specific protein and dsDNA are generally examined by mobility shift assay, involving the use of gel electrophoresis. This method suffers from the lack of quantification and interference of non-specific binding of dsDNA with other proteins. Capillary zone electrophoresis (CZE) can be used to measure specific protein–DNA interactions. The separations achieved by CE and the use of fluorescently labelled DNA probes indicate that this method is capable of effectively resolving dsDNA, ssDNA, complexes of DNA probe with non-specific binding proteins, and complexes with specific Sp1 proteins. Capillary electrophoresis appears to have a high resolving power, and a laser enhanced fluorescence detector provides very sensitive measurements of these molecules.

Acknowledgment

This work was supported by a grant from the National Science Foundation (OSR-9255223).

References

1. Singhal, R. P. (1983). *J. Chromatogr.*, **266**, 359.
2. Bergold, A. and Scouten, W. M. (1983). *Chem. Anal.*, **66**, 149.
3. Weith, H. L., Weibers, J. L., and Gilham, P. T. (1970). *Biochemistry*, **9**, 4396.
4. Ho, N. W. Y., Duncan, R. E., and Gilham, P. T. (1981). *Biochemistry*, **20**, 64.
5. Singhal, R. P., Bajaj, R. K., Buess, C. M., Smoll, D. B., and Vakharia, V. N. (1980). *Anal. Biochem.*, **109**, 1.
6. Hageman, J. H. and Kuehn, G. D. (1977). *Anal. Biochem.*, **80**, 547.
7. Pruess-Schwartz, D., Sebti, S. M., Gilham, P. T., and Baird, W. M. (1984). *Cancer Res.*, **44**, 4104.
8. Annamalai, A. E., Pal, P. K., and Colman, R. F. (1979). *Anal. Biochem.*, **99**, 85.
9. Duncan, R. E. and Gilham, P. T. (1975). *Anal. Biochem.*, **66**, 532.
10. McCutchan, T. F., Gilham, P. T., and Soll, D. (1975). *Nucleic Acids Res.*, **2**, 853.
11. Wilk, H.-E., Kecskemethy, N., and Schafer, K. P. (1982). *Nucleic Acids Res.*, **10**, 7621.
12. Singhal, R. P. and DeSilva, S. S. M. (1992). *Adv. Chromatogr.*, **31**, 293.
13. Lorand, J. P. and Edwards, J. O. (1959). *J. Org. Chem.*, **24**, 769.
14. Barker, S. A., Chopra, A. K., Hatt, B. W., and Somers, P. J. (1973). *Carbohydr. Res.*, **26**, 33.
15. Singhal, R. P., Ramamurthy, B., Govindaraj, N., and Sarwar, Y. (1990). *J. Chromatogr.*, **543**, 17.
16. Aronoff, S., Chen, T., and Cheveldayoff, M. (1975). *Carbohydr. Res.*, **40**, 299.
17. Yurkevich, A. M., Kolodkina, I. I., Ivanova, E. A., and Pichuzhkina, E. I. (1975). *Carbohydr. Res.*, **43**, 215.
18. Boeseken, J. (1949). *Adv. Carbohydr. Chem.*, **4**, 189.
19. Jandera, P. and Churacek, J. (1974). *J. Chromatogr.*, **98**, 55.
20. Singhal, R. P. (1983). *Progr. Nucleic Acids Res. Mol. Biol.*, **28**, 75.
21. Singhal, R. P. and Fallis, P. A. M. (1979). *Progr. Nucleic Acids Res. Mol. Biol.*, **23**, 227.
22. Singhal, R. P. and Vakharia, V. N. (1983). *Nucleic Acids Res.*, **11**, 4257.
23. Jacobson, A. (1987). In *Methods in enzymology* (ed. S. L. Berger and A. R. Kimmel). Academic Press, New York, Vol. 152, p. 254.
24. Xian, J. (1994). Application of capillary electrophoresis for the separation of biomolecules. Doctoral dissertation, Wichita State University, p. 96.
25. Singhal, R. P. and Xian, J. (1996) In *Application of capillary electrophoresis to pharmaceutical and biochemical analysis* (ed. D. M. Radzik and S. M. Lunte). Pergamon Press, New York, p. 387–424.
26. Singhal, R. P., Hughbanks, D., and Xian, J. (1992). *J. Chromatogr.*, **609**, 147.
27. Singhal, R. P. and Xian, J. (1993). *J. Chromatogr.*, **652**, 47.
28. Singhal, R. P., Xian, J., and Otim, O. (1996). *J. Chromatogr.*, **756**, 263–78.
29. Taylor, J. A. and Yeung, E. S. (1992). *Anal. Chem.*, **64**, 1741.
30. Singhal, R. P., Otim, O., and Sheikh, S. S. (1995). American Society of Biochemistry and Molecular Biology–American Chemical Society Meeting, San Francisco, CA, May 1995, Abstract No. 6065, (pp. 21–5).

5

Affinity chromatography of oligosaccharides and glycopeptides

RICHARD D. CUMMINGS

1. Introduction

One of the major obstacles encountered in the study of glycoconjugates is the difficulty in purifying them for detailed structural characterization. Typically, a single glycoprotein contains multiple glycans linked to serine, threonine, asparagine, and other residues. The structures of each glycan may be site specific and there may be significant microheterogeneity in the glycan structures at each site. This problem is compounded when a mixture of glycoproteins, for example those derived from total membranes of a cell or tissue, is the object of study. In these cases, fractionation and purification of glycans by charge and/or size may not be feasible or practical.

A technique that has been exceptionally useful for purifying and analysing mixtures of oligosaccharides is affinity chromatography on immobilized plant and animal lectins. Lectins are carbohydrate-binding proteins that are not immunoglobulins and do not catalyse reactions with the bound sugars. They recognize particular structural elements of glycans with high affinity. Immobilized lectins, in the technique called lectin-affinity chromatography, provide a means of separating oligosaccharides in complex mixtures by virtue of the structural characteristics of each glycan, independent of size and/or charge. Lectin-affinity chromatography is rapid and specific and allows the facile identification and isolation of complex carbohydrates. In this chapter I will attempt to provide a straightforward guide for affinity chromatography of glycans on immobilized lectins. A number of excellent articles have been written in recent years detailing the overall uses of lectins and presenting techniques for the chromatography of oligosaccharides and glycoproteins on columns of immobilized lectins and the reader is encouraged to consult these references for additional information (1–6).

2. Types of lectin used in affinity chromatography

Since Stillmark's discovery of ricin in the seeds of the castor bean plant a little over a hundred years ago, many lectins have been found in a variety of plants, animals, and bacteria. One of the most famous lectins is the plant lectin concanavalin A (Con A). This protein, along with urease, was one of the first proteins originally crystallized by Sumner from jack beans. Con A was later shown to be a lectin with mitogenic and cell-agglutinating activity. The biological effects of lectins on animal cells and the widespread occurrence of lectins in plants led to fruitful searches for lectins in animal tissues in the 1960s and 1970s. The first two lectins discovered in animals were the hepatic asialoglycoprotein receptor and a small, soluble galactose-binding protein, now called galectin-1, found in heart, spleen, and other organs. Nearly a hundred different animal lectins are now known to exist; they are involved in a variety of biological processes, including the recruitment of lymphocytes to endothelium, cell signalling, cell adhesion to extracellular matrix, recognition of viruses and pathogens, and lysosomal targeting.

Because of their inexpensive nature and ease of purification, however, plant lectins are still used more widely than animal or bacterial lectins in studies of glycoconjugates. Some of the commonly used plant and animal lectins are listed in *Figure 1*. Each of the lectins is characterized by high affinity binding to a specific glycan determinant, which is indicated within the boxed area. In addition to these high affinity determinants, some plant lectins may bind weakly to structurally related glycans. Even this weak affinity can be exploited for specific isolation of desired glycans (7). The exquisite specificity of lectins is often overlooked in discussions of lectins that stress their monosaccharide binding specificity. For example, though Con A binds to α-methylmannoside weakly, it displays high affinity binding to high mannose- and hybrid-type *N*-glycans and weaker affinity binding to bi-antennary complex type *N*-glycans. However, Con A does not interact with tri-/tetra-antennary complex-type *N*-glycans or with *O*-glycans. Thus, chromatography on a column of immobilized Con A can facilitate the rapid separation of a mixture of glycans into specific subclasses. The ability to bind complex glycans better than simple mono- and disaccharides is characteristic of lectins and suggests that the binding sites of many lectins can accommodate a relatively large glycan.

These complex determinants recognized by lectins are, however, not generally available to use as haptens for specific elution. Rather, advantage is taken of the fact that lectins weakly recognize simple mono-, di-, and trisaccharides. Relatively high concentrations of these haptens can be used for the specific elution of bound glycans from a column of immobilized lectin, as shown in *Table 1*. For some lectins, such as CD22, no simple hapten is available; glycans are passed over the lectin at low temperature (4 °C) and elution of interactive glycans is accomplished by a temperature jump that increases the exchange rate of the lectin (8).

Table 1. Hapten elution of glycans bound by immobilized lectins

Lectin[a]	**Recommended coupling density for chromatography of free glycans or glycopeptides (mg/ml)**	**Method of elution**
Con A	10–15	α-methylglucoside (10 mM) (bi-antennary *N*-glycans) α-methylmannoside (100 mM) (hybrid and high-mannose-type *N*-glycans)
Pea lectin	5–10	α-methylmannoside (100 mM)
Lentil lectin	5–10	α-methylmannoside (100 mM)
L4-PHA	3–5	Gal*N*Ac (0.4 M)
E4-PHA	3–5	Gal*N*Ac (0.4 M)
RCA-I	5–20	lactose (100 mM)
WGA	3–10	Glc*N*Ac (50 mM)
TPA[b]	6–15	fucose (0.2 M)
Tomato lectin	5–10	chitotriose/chitotetraose mixture (10 mg/ml)
GSI-B4	5–15	raffinose (10 mM)
Jacalin lectin	3–5	α-methylgalactoside (50 mM)
MAL	3–5	lactose (50 mM)
SNA	3–5	lactose (50 mM)
HPA	3–5	Gal*N*Ac (25 mM)
AAL	8–10	fucose (5 mM)
PWM	5–10	chitotriose/chitotetraose mixture (10 mg/ml)
DSA	5–10	chitotriose/chitotetraose mixture (10 mg/ml)
UEA-I	3–5	fucose (10 mM)
WFA	3–5	Gal*N*Ac (50 mM)
Mannose-6-phosphate receptor	3–5	mannose-6-phosphate (10 mM)
Galectin-1	3–5	lactose (50 mM)
CD22	2–5	temperature shift from 4°C to 25°C

[a]Full names of lectins are given in Figure 1.
[b]*Tetragonolobus purpureas* agglutinin.

3. Methods of coupling lectins

3.1 Covalent methods

The preferred method for affinity chromatography of glycans on immobilized lectins is to conjugate the lectins covalently to an insoluble matrix. Several matrices with different functional chemistries are available (*Table 2*). The selection of specific supports often depends on convenience, expense, and

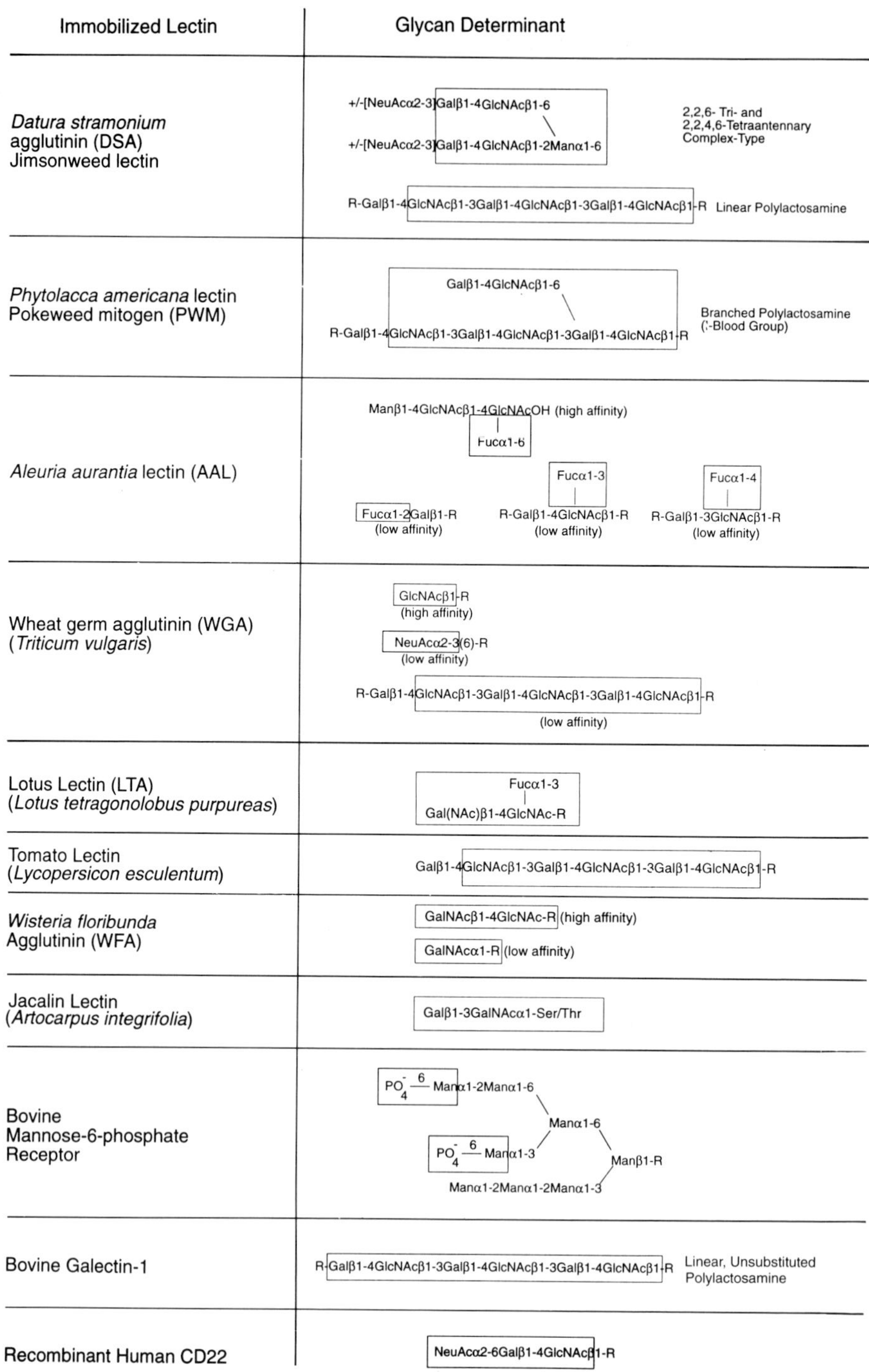

Figure 1. Glycan determinants recognized by lectins. The boxed area within each glycan structure is known to be necessary for high affinity interactions of the glycan with the designated lectin.

Immobilized Lectin	Glycan Determinant
Lentil Lectin (LCA) (*Lens culinaris*) and Pea Lectin (PSA) (*Pisum sativum*)	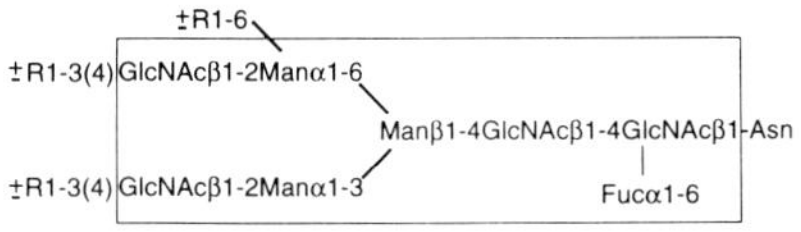 ±R1-6; ±R1-3(4) GlcNAcβ1-2Manα1-6; Manβ1-4GlcNAcβ1-4GlcNAcβ1-Asn; ±R1-3(4) GlcNAcβ1-2Manα1-3; Fucα1-6 — Bi-/Triantennary Complex-Type with Core Fucose
Leukoagglutinating PHA (L4-PHA) (*Phaseolus vulgaris*)	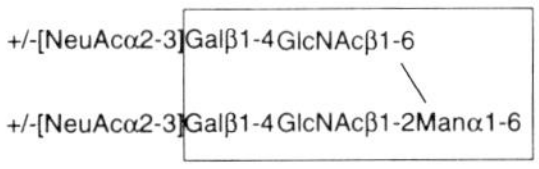 +/-[NeuAcα2-3]Galβ1-4GlcNAcβ1-6; +/-[NeuAcα2-3]Galβ1-4GlcNAcβ1-2Manα1-6 — 2,2,6-Tri- and 2,2,4,6-Tetraantennary Complex-Type
Concanavalin A (Con A) (*Canavalia ensiformis*)	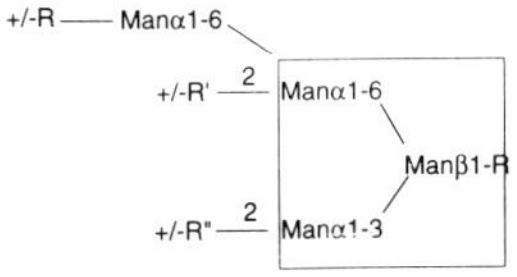 +/-R — Manα1-6; +/-R' —2— Manα1-6; Manβ1-R; +/-R" —2— Manα1-3 — High Mannose-Type, Hybrid-Type, Biantennary Complex-Type
Erythroagglutinating PHA (E4-PHA) (*Phaseolus vulgaris*)	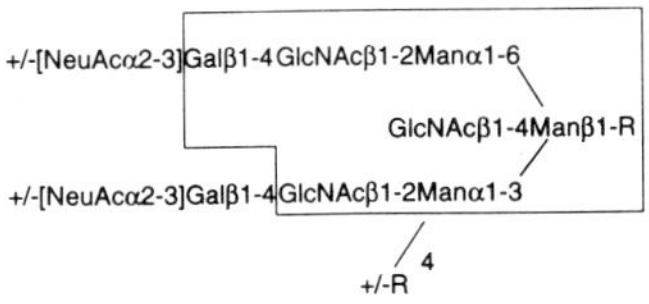 +/-[NeuAcα2-3]Galβ1-4GlcNAcβ1-2Manα1-6; GlcNAcβ1-4Manβ1-R; +/-[NeuAcα2-3]Galβ1-4GlcNAcβ1-2Manα1-3; 4 +/-R — Bisected Bi-/Triantennary Complex-Type
Ricinus communis Agglutinin-I (RCA-I)	Galβ1-4GlcNAcβ1-R
Maackia amurensis Leukoagglutinin (MAL)	NeuAcα2-3Galβ1-4GlcNAcβ1-R
Sambucus nigra Agglutinin (SNA)	NeuAcα2-6Gal(NAc)-R
Ulex europaeus Agglutinin-I (UEA-I)	Fucα1-2Gal-R
Griffonia simplicifolia Agglutinin-I-B4 (GSI-B4)	Galα1-3Gal-R
Helix pomatia agglutinin (HPA) Snail lectin	GalNAcα1-R

Table 2. Affinity matrices for coupling lectins for affinity chromatography

Name	Vendor	Matrix	Functional group
Emphaze™	Pierce	cross-linked bis-acrylamide	azlactone
Affi-Gel-10	Bio-Rad	cross-linked agarose	*N*-hydroxysuccinimide
Affi-Prep-10	Bio-Rad	acrylic polymer	hydrazide
CNBr-activated Sepharose 6MB	Pharmacia	cross-linked agarose	CNBr
Formyl-agarose	Seikagaku Kogyo	cross-linked agarose	formyl

availability. The three most commonly used supports are Emphaze™, Affi-Gel, and CNBr-Sepharose. Emphaze™ beads with azlactone chemistry are convenient for many reasons. The azlactone group is more stable in water than *N*-hydroxysuccinimide; coupling is rapid and quantitative within 2–3 h; Emphaze™ beads have a very high coupling capacity (~ 30 mg/ml), a very low non-specific binding, and very low compressibility; they provide even flow rates.

The coupling of *Tetragonolobus purpureas* agglutinin (TPA, lotus lectin) to Emphaze™ is described in *Protocol 1*. In this type of coupling, as for all lectin chemical coupling methods, a hapten sugar is added to the coupling reaction to partly protect the lectin from undesirable inactivation. Coupling is routinely 90% or greater with this resin and can be readily estimated by direct protein determination by absorbance at 280 nm of the reaction buffer. Most immobilized lectins display remarkable stability. When stored at 4°C in 0.02% (w/v) sodium azide to prevent bacterial growth, most immobilized lectins remain stable and active for up to 6 months or longer, depending on the frequency of use and conditions of samples analysed.

Protocol 1. Conjugation of *Tetragonolobus purpureas* agglutinin (TPA) to Emphaze™ beads

Equipment and reagents

- Phosphate/citrate buffer: 100 ml of 0.05 M sodium phosphate, 0.6 M sodium citrate (pH 7.5) containing 0.2 M L-fucose (Sigma F2252)
- Tris buffered saline/azide (TBS/NaN_3): 10 mM Tris, 150 mM NaCl, 1 mM $CaCl_2$, 1 mM $MgCl_2$, 0.02% NaN_3 (pH 8.0)
- NaCl
- $CaCl_2$
- $MgCl_2$
- TPA agglutinin (TPA, 'lotus lectin'; Sigma L9254)
- Emphaze™ beads (Pierce 53110)
- 2-aminoethanol (ethanolamine) (Sigma E9508), 3.0 M in water
- 3 ml chromatography column (Bio-Rad 737-0516 or equivalent)

Method

1. Dissolve 10 mg of TPA in 2 ml of phosphate/citrate buffer in a 15 ml plastic disposable tube.

2. Add 187 mg of dry Emphaze™ powder to the TPA solution and cap the tube.
3. Rotate the tube for 3–4 h at room temperature.
4. Stop the coupling by adding 0.5 ml of 3.0 M ethanolamine (pH 9.0)
5. Rotate the tube for an additional 3 h at room temperature.
6. Let the beads settle and remove supernatant. Save the supernatant and determine the amount of protein by absorbance at 280 nm or by common chemical assays for protein. The amount of coupled protein is taken as the total starting amount minus that recovered in the supernatant following coupling.
7. Wash the beads in the tube three times by adding 5 ml of TBS/NaN_3.
8. Transfer the beads to a small chromatography column. Alternatively, a 1 ml plastic disposable pipette may be used as a convenient column, as described in *Protocol 2*.
9. Wash and equilibrate column in TBS/NaN_3.[a]

[a]Note: A variety of buffers can be used interchangeably with these columns. For example, PBS/NaN_3, as described in *Protocol 2,* is also routinely used.

Affi-Gel-10 and -15 (Bio-Rad) are composed of cross-linked agarose containing *N*-hydroxysuccinimide functional groups. Coupling of lectins to these supports is easy and efficient. However, the *N*-hydroxysuccinimide ester hydrolyses quickly in water. (Partly for this reason, the shelf-life of opened bottles of Affi-Gel is unpredictable unless the residual gel is degassed.) When using Affi-Gel for coupling, care should be taken to wash the resin quickly with buffer prior to addition to the coupling reaction. Lectins coupled to Affi-Gel are relatively stable and leaching is not generally a problem. CNBr–Sepharose comes as a dry powder to which protein can be readily coupled through the one-carbon spacer involving an iso-urea linkage. However, coupled proteins tend to leach from the gel due to the inherent instability of the linkage and the matrix has undesirable ion-exchange properties. Proteins can be coupled to formyl-agarose by the reductive amination method. In this approach the amino groups of proteins form a Schiff base with the formyl groups and this is reduced by sodium cyanoborohydride. The protocols for coupling to these other supports are very similar to those involving Emphaze™, except that the buffer concentrations can be lowered to 0.1–0.2 M. Protective haptens are included as described in *Protocol 1*. For these other supports, it is necessary to read the manufacturer's instructions carefully to ensure optimal yields. Coupling densities with these other supports cannot be determined by direct protein assays on the reaction supernatant, due to interfering by-products. To determine the amount of uncoupled lectin, the lectin in reaction buffer is precipitated at 4°C by addition of trichloroacetic acid to a final concentration of 10% (w/v). The precipitated protein can then be resuspended in water or an appropriate buffer for direct protein determinations.

Some lectins are available commercially as conjugated supports on agarose, Sepharose, etc. In these cases it is important to determine that the coupling density of the lectin is in the range useful for chromatography of glycans (see *Table 1*) and that the immobilized lectin is active. In some cases this information may be available from the manufacturer, but often the investigator must make the determination. For this, it is necessary to have a variety of standard glycans known to interact with the particular lectin under testing. Several companies, including Sigma, Oxford Glycosciences, Dextra Laboratories, GlycoTech, and V-Labs, now sell purified glycans useful for standardization of lectin columns.

There are a variety of methods for 'tagging' glycans, including both metabolic and non-metabolic approaches. The metabolic approach requires that glycans be prepared from living cells or tissues grown in the presence of radioactive precursors, usually ^{3}H- or ^{14}C-labelled monosaccharides. But $^{35}SO^{2-}{}_4$ and $^{32}PO^{3-}{}_4$ have also been used to tag sulfated and phosphorylated glycans, respectively.

In non-metabolic radiolabelling glycans may be labelled at the non-reducing end by treatment with either $NaIO_4$ or galactose oxidase, followed by $NaB[^{3}H]_4$ reduction (9). Alternatively, free reducing glycans can be labelled directly by reduction with $NaB[^{3}H]_4$ (9).

It is also possible to re-acetylate *N*-glycans isolated by hydrazinolysis procedures with ^{14}C-labelled acetic anhydride (10). In addition, both glycosyltransferases and glycosidases have been used in a variety of approaches to radiolabel glycans, but, by necessity, these methods require a modification of the fundamental sequence of the glycan (11,12).

There are several direct chemical methods for detecting sugars, such as the phenol–sulfuric acid assay and the ferricyanide reducing sugar assay (13,14). Fluorescently labelled glycans have also been prepared from glycopeptides and free glycans (15,16).

3.2 Non-covalent methods

It is also possible to couple lectins to supports by non-covalent methods. An example of this method is the binding of one lectin by another. *Griffonia simplicifolia* agglutinin IB4 (GSI-B4) is a glycoprotein with high mannose-type *N*-glycans that is bound with high affinity by Con A–Sepharose (17). Glycans that themselves do not bind to Con A–Sepharose can then be passed over this column of GSI-B4–Con A–Sepharose. Glycans with terminal α-linked galactosyl residues are bound by the GSI-B4 and can be eluted with specific haptens without disturbing the 'sandwich' nature of the support. It is conceivable that biotinylated lectins could be bound by immobilized streptavidin/avidin to give a type of noncovalent sandwich. In these cases the advantages are that no chemical derivatization is involved and there is little or no inactivation of the lectin. Furthermore, coupling is quantitative and can be adjusted to give precisely the desired amount of immobilized lectin. A disadvantage of this approach is that the lectins can leach from the column over time and the column has a relatively short half-life compared with covalently coupled lectins.

4. Critical factors for successful lectin-affinity chromatography

A number of factors must be considered for successful chromatography of oligosaccharides and glycopeptides on immobilized lectins; these include:

- lectin density
- activity of immobilized lectin
- column flow rate
- column temperature
- column dimensions
- buffer composition

Free oligosaccharides can be prepared from intact glycoproteins by a variety of enzymatic or chemical methods. The enzymatic methods utilize endoglycosidases, such as peptide:*N*-glycosidase F (PNGaseF), which cleaves between the asparagine and Glc*N*Ac residue in all high mannose-, hybrid-, and complex-type *N*-glycans. *O*-glycanase is an endoglycosidase useful for *O*-glycans, but it has a very restricted specificity and only releases the disaccharide Galβ1–3Gal*N*Ac from its attachment to serine/threonine residues in glycopeptides and glycoproteins. Hydrazinolysis is the preferred method for chemical release of intact *N*- and *O*-glycans and involves a base-catalysed, anhydrous cleavage of *O*- and *N*-glycosides. In addition, treatment of glycopeptides and glycoproteins with mild base/borohydride effects the beta-elimination of *O*-glycans from serine/threonine residues. An alternative method is to prepare glycopeptides from intact proteins, by treatment with either trypsin or a mixed protease, such as Pronase or proteinase K. The resultant glycopeptides generally have only one glycan. Ceramidases are now available to cleave the ceramide from glycosphingolipids and the released glycans can then be analysed by lectin affinity chromatography. In most cases the fractionation of free oligosaccharides is similar to that of glycopeptides. However, lectins which recognize the core α1,6-linked fucose in complex-type *N*-glycans bind differentially to glycopeptides versus free oligosaccharides. *Aleuria aurantia* lectin (AAL) binds best to the core α1,6-linked fucose in free *N*-glycans that have been released from peptide and reduced. In contrast, pea and lentil lectins, which recognize certain *N*-glycans containing a core α1,6-linked fucose residue, bind to glycopeptides better than to free, reduced oligosaccharides.

The density of lectin coupling required is often not fully appreciated in lectin affinity chromatography. For example, *Ricinus communis* agglutinin I (RCA-I)-agarose binds glycans containing a single terminal Gal residue in β1,4-linkage. However, optimal interactions require a density of 20 mg/ml of lectin (18). Such high density RCA-I is not commercially available and must be prepared by the investigator. The same is true of lotus lectin (TPA), which also requires high density coupling for maximum interactions with fucosylated oligosaccharides (19a). The recommended coupling densities for lectins are shown in *Table 1*. The higher density coupling is preferred in most cases when free glycans or glycopeptides are being analysed. However, with intact glycoproteins and glycopeptides containing multiple glycans, the avidity of binding to an immobilized lectin can be so high that hapten sugars cause the bound material to leach off and there is usually poor recovery.

Many plant lectins bind relatively slowly to glycans and for this reason it is important to allow sufficient time for interaction between the glycans and the lectins. This can be achieved by having a very slow flow rate in the range of 0.1–0.5 ml/min and by altering the column dimensions. Column geometry is a critical factor and columns with length:width ratio of ~ 10:1 or greater provide the optimal separation of glycans (2). Ideally, chromatography should be performed at low temperature, e.g. between 4 and 10°C. The lower temperature reduces the off-rate of glycans from lectins and helps preserve the activity of the lectin over time. However, due to the inexpensive nature of plant lectins and the inconvenience of low temperature chromatography, most investigators perform this chromatography at room temperature. The lectin columns should be stored routinely at 4°C, and allowed to warm up to room temperature about 1–2 h before use. When used at room temperature, it is important to keep the buffers at the same temperature as the column to avoid air bubbles being trapped in the columns.

Some plant lectins are metalloproteins, containing bound Mn^{2+}, Mg^{2+}, and/or Ca^{2+} (19). For this reason, it is important to include these cations at 1 mM concentrations in all buffers to prevent slow inactivation of the immobilized lectin.

5. Chromatography of glycans on a column of immobilized lectin

The common technique of lectin affinity chromatography employs gravity flow columns or columns pumped under low pressure. Although some lectins have been used successfully in HPLC supports, the simplicity and ease of use of small lectin columns is attractive. However, the investigator should ensure that the amount of immobilized lectin is sufficient to bind quantitatively all appropriate glycans. To test this, it is necessary to pass samples over a column and recover the unbound material. After eluting the bound glycans the column is regenerated and the unbound material passed back over the column. If additional glycans bind, then the investigator can gauge the approximate capacity of the column for glycans and either increase the size of the column (without changing the coupling density) or simply repeat the chromatography with the unbound material until all appropriate glycans recognized by the lectin are depleted. As an approximation, a column containing 10 mg/ml of a typical plant lectin (with a subunit size of ~ 25 kDa) can bind ~ 50 nmol of glycans. This number may be considerably lower if a high percentage of the lectin is immobilized or the flow conditions and column geometry are inappropriate. Ideally, a standard glycan known to interact with a specific lectin should be passed over the column to ensure that the lectin is active and the separation is adequate.

The technique of lectin-affinity chromatography was developed using relatively small amounts of samples in the range of less than 25 nmol. If larger quantities of samples are available, it is possible to scale up the amount of immobilized lectin used and use proportionally larger columns.

Space does not permit a protocol to be written for each lectin listed in *Figure 1* and *Table 1*. However, with small amounts of sample in the range of less than 25 nmol, the basic chromatographic conditions are similar and the only variability with different lectins is the differential degree of coupling and the conditions and haptens required for elution of interactive glycans. An illustration of the isolation of oligosaccharides containing polylactosamine chains on tomato lectin–Sepharose (see *Figure 1*) is given in *Protocol 2*.

Protocol 2. Isolation of polylactosamine-containing glycans on tomato lectin–Sepharose

Equipment and reagents

- Tomato lectin–Sepharose (1 ml containing 3.5 mg/ml lectin) (prepared by conjugation of tomato lectin to CNBr-Sepharose 6MB Pharmacia) (see refs 20,21)
- Mixture of chitotriose and chitotetraose (prepared according to ref. 22)
- Phosphate-buffered saline with azide (PBS/NaN_3): 6.7 mM KH_2PO_4, 0.15 M NaCl (pH 7.4), containing 0.02% NaN_3. Prepare fresh (100 ml).
- 1 ml plastic disposable pipette (Fisher, 13-678-11B)
- Glass wool
- PVC tubing ¼ inch i.d. (Fisher 14-176-218)
- Sample of oligosaccharides[a]
- Radioactivity counter set for ^3H detection

Method

1. Prepare a 3 ml solution of chitotriose/chitotetraose (10 mg/ml) in PBS/NaN_3.
2. Resuspend the tomato lectin–Sepharose in 2 ml of PBS/NaN_3.

Protocol 2. *Continued*

3. Remove the cotton plug from the plastic pipette. Place a small wad of glass wool in the top of the pipette and push the glass wool to the bottom of the pipette with a small wire. Attach narrow PVC tubing as an outlet. Clamp the column in place on a ring stand.
4. While gently tapping the outlet tube of the column, add PBS/NaN_3 to about two-thirds of the volume of the column. Begin to add the slurry of tomato lectin–Sepharose with a 9-inch glass Pasteur pipette, while letting the buffer drain slowly. Continue adding until all the tomato lectin–Sepharose has been placed in the column to a total volume of approximately 1 ml.
5. Suspend the sample in 0.5 ml of PBS/NaN_3. Collect 1 ml fractions, either by hand or with an automatic fraction collector. Use an additional 0.5 ml of PBS/NaN_3 to rinse the tube containing the sample and apply this to the column. Elute the column with PBS/NaN_3. This can be accomplished by repeatedly adding 0.5 ml of PBS/NaN_3 to the column or setting up a reservoir on top of the column with a 10 ml syringe connected to the column by PVC tubing.
6. After collection of fraction 10, begin adding the hapten solution, which is a mixture of chitotriose/chitotetraose (10 mg/ml).
7. After addition of the hapten solution, regenerate the column with 25 ml of PBS/NaN_3.
8. Remove aliquots of each fraction in the wash and the hapten elution and test for distribution of the sample. The polylactosamine-containing glycans will be bound and eluted by the hapten. These bound glycans will have a repeating Gal–Glc*N*Ac motif as shown in *Figure 1*.

[a]The sample may be ^{3}H-labelled and at least 1000 c.p.m. should be analysed. Overall, the sample should contain no more than 100 nmol of total polylactosamine-containing glycans.

After successful separation of glycans on a column of immobilized lectin, it may be necessary to remove the hapten sugars used for elution. In this case, the samples are dried by lyophilization, resuspended in a small volume of water and 'desalted' by passage over a column (1 cm × 50 cm) of Sephadex G-25 or Bio-Gel P-2 in 7% propan-1-ol in water. (The propan-2-ol serves to prevent bacterial contamination of the columns.) The glycopeptides or oligosaccharides are recovered in the fractions containing the void volume and the hapten sugars are recovered in the included fractions.

6. Serial lectin-affinity chromatography

The combined use of many lectins for the sequential or serial fractionation of a mixture of glycans is termed serial lectin-affinity chromatography or SLAC

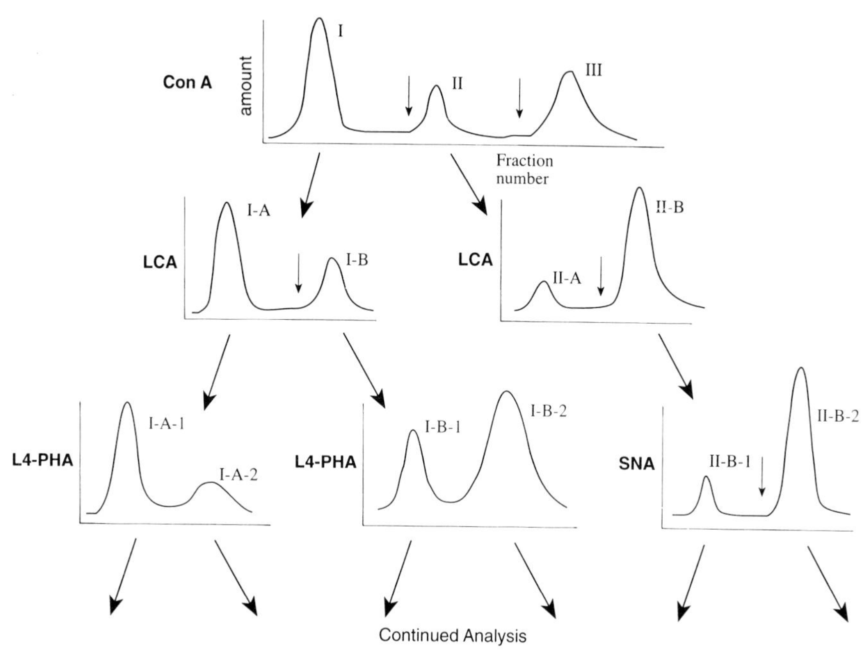

Figure 2. Serial lectin-affinity chromatography. The combined use of several lectins is illustrated for the fractionation of a complex mixture containing many of the glycans listed in *Figure 1*. In this example, the initial chromatography is performed on ConA–Sepharose and glycans are differently eluted by hapten sugars as illustrated by the arrows. The glycans designated I, II, and II are collected and concentrated and the hapten sugars are removed by size-exclusion column chromatography. The glycans are then serially passed over either immobilized lentil lectin (LCA), L4-PHA, or SNA. The partly purified glycans can then be further fractionated either over additional lectins or by other conventional procedures, such as high-performance liquid chromatography.

(23). This approach can be used to purify the majority of glycans in complex mixtures in many cases. The predominant use of SLAC is for the separation of *N*-glycans. Few lectins are known to interact specifically with *O*-glycans or with glycolipid-derived glycans. However, many lectins recognize terminal sugars in specific linkages that may occur on both *N*- and *O*-glycans, and even on glycolipid-derived glycans and free oligosaccharides found in body fluids, e.g. milk or urine.

The general scheme of SLAC is to use Con A–Sepharose to remove hybrid/high-mannose-type *N*-glycans and bi-antennary complex-type *N*-glycans by differential elution, as shown in *Figure 2*. The hybrid/high mannose-type *N*-glycans and bi-antennary *N*-glycans require 100 mM α-methylmannoside and 10 mM α-methylglucoside, respectively, for elution. The unbound material is designated fraction I, and the bound samples are designated fractions II and III. These samples may then be passed over another lectin, e.g. leukoagglutinating phytohaemagglutinin (L4-PHA)–agarose, for the isolation of specifically branched glycans (*Figures 1* and *2*). The serial chromatography of glycans on several lectins can proceed until the mixture has been sufficiently fractionated. The partly purified glycans can then be further purified by other chromatographic procedures, such as high-performance liquid chromatography, size-exclusion column chromatography, and paper chromatography/electrophoresis.

Protocol 3. Serial lectin-affinity chromatography of *N*-glycans on immobilized Con A, lentil lectin (LCA), and L4-PHA[a]

Equipment and reagents

- Con A–Sepharose (2.0 ml containing 15 mg/ml) (it is recommended to purchase this conjugate from Pharmacia)
- LCA–Sepharose (2.0 ml containing 10 mg/ml) prepared by conjugation to CNBr–Sepharose according to the manufacturer's instructions in the presence of 100 mM α-methylmannoside)
- L4-PHA–agarose (5.0 ml containing 5 mg/ml) prepared by conjugation to Affi-Gel-10 according to the manufacturer's instructions in the presence of 0.4 M Gal*N*Ac)
- TBS/NaN_3: 10 mM Tris–HCl, 150 mM NaCl, 1 mM $CaCl_2$, 1 mM $MgCl_2$, pH 8.0, containing 0.02% (w/v) NaN_3. Prepare fresh.
- PBS/NaN_3 prepared according to *Protocol 2*
- α-Methylmannoside (Sigma M6882), 100 ml of a 10 mM solution in TBS/NaN_3; store at 4°C and warm to room temperature before use
- Sephadex G-25 medium (Sigma 29,328-8)
- α-Methylglucoside (Sigma M9376), 100 ml of a 10 mM solution in TBS/NaN_3; store at 4°C and warm to room temperature before use
- *N*-Acetylgalactosamine (Gal*N*Ac) (Sigma A2795), 100 ml of a 0.4 M solution in PBS/NaN_3; store at 4°C and warm to room temperature before use
- Propan-1-ol (Sigma 29,328-8), 1 L of 7% (v/v) solution freshly prepared with deionized water
- Sample of oligosaccharides (the sample may be radiolabelled, as discussed in *Protocol 2*, or tagged in another way (see step 10b of this protocol)). The molecular weight of the glycans should be >1000.
- Two 3 ml chromatography columns (Bio-Rad 727-0516) (0.5 cm × 15 cm)
- 6 ml chromatography column (Bio-Rad 737-0716) (0.7 cm × 15 cm)
- 40 ml chromatography column (Bio-Rad 737-0516) (1 cm × 50 cm)

Method

1. In the 3 ml column place 2 ml of Con A–Sepharose and wash with 20 ml of TBS/NaN_3. Keep at room temperature. This column will not be re-used.
2. In a second 3 ml column place 2 ml of LCA–Sepharose and wash with 20 ml of TBS/NaN_3. Store at 4°C until ready for use and warm to room temperature. This column will be re-used.
3. In the 6 ml column place 5 ml of L4-PHA–agarose and wash with 30 ml of PBS/NaN_3. Store at 4°C until ready for use and warm to room temperature. This column will be re-used.
4. Swell 10 g of Sephadex G-25 in 7% propan-1-ol either overnight or by heating to 60°C for 2 h, followed by cooling to room temperature.
5. In the 40 ml column place approximately 40 ml of Sephadex G-25 and wash the column with 100 ml of 7% propan-1-ol. (This column

will be used to separate oligosaccharides from hapten sugars and buffer salts and will be re-used often. It is stored at room temperature in the presence of 7% propan-1-ol to prevent contamination by micro-organisms.)

6. Suspend the sample of oligosaccharides in 1.0 ml of TBS/NaN_3.
7. Apply the sample to the Con A–Sepharose, while collecting 2.0 ml fractions. Elute with TBS/NaN_3 at a flow rate of 0.5 ml/min and collect 20 fractions.
8. Begin elution with 10 mM α-methylglucoside and collect 20 fractions.
9. Begin elution with 100 mM α-methylmannoside preheated to 60 °C. (This heated solution allows efficient elution of tightly bound glycans and is used instead of a 0.5 M α-methylmannoside solution at room temperature, thus saving on costs and handling in subsequent steps.)
10. (a) If the sample is radiolabelled, remove a 1–5% aliquot of each fraction and count in a liquid scintillation counter. The fractions containing unbound glycans are pooled and designated as in *Figure 2*. Fractions eluted with α-methylglucoside and α-methylmannoside are pooled and designated II and III, respectively.
    (b) Alternatively, the glycans may not be radiolabelled, but may be tagged in some other way, e.g. fluorescently labelled (24) or biotinylated at the reducing end. If the glycans are fluorescently labelled, each fraction may be tested for fluorescence and appropriate fractions pooled. Some glycopeptides may be detectable by absorbance at 214 or 280 nm. If the glycans are not detectable by these methods, it may be necessary to detect them by chemical methods for sugar analysis. However, these methods will not be useful until the glycans are separated from hapten sugars by chromatography on Sephadex G-25, as discussed below.
11. Concentrate the glycans in fractions I, II, and III by either of two methods:
    (a) Either freeze in dry ice–acetone or liquid N_2 and freeze dry.
    (b) Alternatively, concentrate the glycans by shaking or rotating the samples while heating to approximately 50 °C *in vacuo* with a vacuum trap and pump.
12. Resuspend the dried samples in 1.0 ml of 7% propan-1-ol and apply directly to the column of Sephadex G-25. Collect 1.0 ml fractions at a flow rate of 1.0 ml/min and wash the column with 50 ml of 7% propan-1-ol. Glycans may be detected as discussed in step 10. Glycans should elute in fractions 12–20. Pool the desalted glycans and dry as in step 11.

Protocol 3. *Continued*

13. Resuspend the glycans in fractions I and II separately in 1.0 ml of TBS/NaN_3. Resuspend the glycans in fraction III in water or store dry at −20°C.
14. Apply the glycans in fractions I and II separately to the column of LCA–Sepharose, while collecting 2.0 ml fractions at a flow rate of 0.5 ml/min. Collect 20 fractions.
15. Begin elution with 100 mM α-methylmannoside at room temperature and collect 20 fractions.
16. Determine the elution of glycans, as in step 10.
17. Pool the fractions containing glycans not bound by the lectin, either IA or IIA (see *Figure 2*), and those that are bound, either IB or IIB.
18. Concentrate the glycans and desalt on Sephadex G-25, as described in steps 11 and 12.
19. Resuspend the glycans in IA and IB separately in 0.5 ml of PBS/NaN_3 and apply them separately to the L4-PHA–agarose at room temperature, while collecting 1.0 ml fractions. Elute with PBS/NaN_3 at a flow rate of 0.5 ml/min and collect 25 fractions.
20. Begin elution with 0.4 M Gal*N*Ac in PBS/NaN_3, and collect 15 fractions.
21. Detect the glycans as in step 10. Pool, concentrate, and desalt the glycans as in steps 11 and 12. The unbound glycans are designated IA1 and IB1 and the bound or retarded glycans are designated IA2 or IB2. Some glycans do not bind so tightly to L4-PHA and are simply retarded in their elution from the column, as illustrated in *Figure 2*.

For L4-PHA–agarose and all the lectins used in serial lectin-affinity chromatography, the elution of unbound glycans can be compared with that of a standard, such as [^{3}H]ribitol or any sugar known not to bind to the lectin. The glycans in fractions IA1, IB1, IA2, IB2, IIA, IIB, and II may be further fractionated on columns of other lectins illustrated in *Table 1* and *Figure 1*.

[a] In this protocol a sample containing a mixture of oligosaccharides, including *N*-glycans, is sequentially analysed by chromatography on columns of Con A–Sepharose, LCA–Sepharose, and L4-PHA–agarose, as illustrated in *Figure 2*.

Acknowledgements

The work of the author has been supported over the years by a grant (CA37626) from the National Institutes of Health.

References

1. Osawa, T. and Tsuji, T. (1987). *Annu. Rev. Biochem.*, **56,** 21.
2. Merkle, R. K. and Cummings, R. D. (1987). In *Methods in enzymology,* Vol. 138 (ed. V. Ginsburg), p. 232. Academic Press, New York.
3. Kobata, A. and Yamashita, K. (1993). In *Glycobiology: A practical approach* (ed. M. Fukuda and A. Kobata), p. 103. IRL Press, Oxford.
4. Cummings, R. D. (1994). In *Methods in enzymology.* Vol. 230 (ed. W. J. Lennarz and G. W. Hart), p. 66. Academic Press, New York.
5. West, I. and Goldring, O. (1994). *Mol. Biotechnol.*, **2,** 147.
6. Yamamoto, K., Tsuji, T., and Osawa, T. (1995). *Mol. Biotechnol.*, **3,** 25.
7. Renkonen, O., Helin, J., Penttilä, L., Maaheimo, H., Niemelä, R., Leppänen, A., Seppo, A., and Hård, K. (1991). *Glycoconjugate J.*, **8,** 361.
8. Powell, L., Sgroi, D., Sjoberg, E. R., Stamenkovic, I., and Varki, A. (1993). *J. Biol. Chem.*, **268,** 7019.
9. Gamhberg, C. G. and Tolvanen, M. (1994). In *Methods in enzymology*, Vol. 230 (ed. W. J. Lennarz and G. W. Hart), p. 32. Academic Press, New York.
10. Amano, J. and Kobata, A. (1980). In *Methods in enzymology*, Vol. 179 (ed. V. Ginsburg), p. 261. Academic Press, New York.
11. Spillmann, D. and Finne, J. (1989). In *Methods in enzymology*, Vol. 179 (ed. V. Ginsburg), p. 270. Academic Press, New York.
12. Whiteheart, S. W., Passaniti, A., Reichner, J. S., Holt, G. D., Haltiwanger, R. S., and Hart, G. W. (1982). In *Methods in enzymology*, Vol. 179 (ed. V. Ginsburg), p. 82. Academic Press, New York.
13. Dubois, M., Giles, K. A., Hamilton, J. K., Rebers, P. A., and Smith, F. (1956). *Anal. Chem.*, **28,** 350.
14. Park, J. T. and Johnson, M. J. (1949). *J. Biol. Chem.*, **181,** 149.
15. Rice, K. G. (1994). In *Methods in enzymology*, Vol. 247 (ed. Lee), p. 30. Academic Press, New York.
16. Tamura, T., Wadhwa, M. S., Chiu, M. H., Cooradi da Silva, M. J., McBroom, T., and Rice, K. G. (1994). In *Methods in enzymology*, Vol. 247 (ed. Lee), p. 43. Academic Press, New York.
17. Wang, W.-C., Clark, G. F., Smith, D. F., and Cummings, R. D. (1988). *Anal. Biochem.*, **175,** 390.
18. Rivera-Marrero, C. A. and Cummings, R. D. (1990). *Mol. Biochem. Parasitol.*, **43,** 59.
19. Lis, H. and Sharon, N. (1986). *Annu. Rev. Biochem.*, **55,** 35.

19a. Yan, L., Wilkins, P. P., Alvarez-Manilla, G., Do, S.-I., Smith, D. F. and Cummings, R. D. (1997). *Glycoconjugate J.* (In Press).

20. Merkle, R. K. and Cummings, R. D. (1987). *J. Biol. Chem.*, **262,** 8179.
21. Merkle, R. K. and Cummings, R. D. (1987). *Plant Sci.*, **48,** 71.
22. Rupley, J. A. (1964). *Biochim. Biophys. Acta*, **83,** 245.
23. Cummings, R. D. and Kornfeld, S. (1982). *J. Biol. Chem.*, **257,** 11235.
24. Hase, S., Ibuki, T., and Ikenaka, T. (1984). *J. Biochem.*, **95,** 197.

6

Immunoaffinity and IgG receptor technologies

MIGUEL A. J. GODFREY

1. Introduction

Historically, affinity chromatographic techniques have exploited the presence of naturally occurring biospecific interactions found in nature. Perhaps the most useful of these to biomedical science is the binding of antibodies to antigens, especially as the presence and binding specificities of antibodies may be manipulated by selective operator intervention. This chapter aims to introduce the laboratory scientist to affinity chromatographic methods of purifying antibodies and to demonstrate the use of immobilized antibodies as a tool in preparative affinity chromatography.

2. Antibodies

Antibodies are molecules found in the serum and secretions of animals; they are usually produced in response to, and are specific for, foreign substances or *antigens*. Antibodies form the group of proteins referred to as immunoglobulins with molecular weights ranging from 150 000 to 900 000, a carbohydrate content in the range 2–14%, and isoelectric points (p*I*) in the range 4.4–9.5. The production of immunoglobulins may be stimulated by *immunization*, introducing immunogens (purified antigen preparations) of molecular weight 5000 or greater to an animal's immune system (1). The production of antibodies in an animal by immunization has been widely reported and has been comprehensively covered by two other books in this series (2,3).

The resulting antibodies isolated from blood samples of immunized animals are polyclonal in nature, in that they consist of a heterogeneous population of immunoglobulins, a few clones of which will be directed towards the antigenic sites on the immunogen (*epitopes*). The antiserum thus produced contains antibodies secreted from a number of cells and will be unique to the animal itself; the supply of similar antisera will be limited to the lifetime of the animal. Monoclonal antibodies (mAbs) are a homogeneous immunoglobulin population and are secreted by clones of a cell *constructed* by the fusion of an immortal myeloma cell with an antibody secreting cell (splenocyte or lymphocyte) from an immunized animal to produce a hybridoma cell (4). Hybridoma construction is generally carried out using splenocytes isolated from a hyperimmunized mouse and results in an immortalized cell line secreting a murine mAb of consistent characteristics. Cell lines secreting non-murine antibodies, such as ovine, bovine, and humanized mAb, may be constructed by interspecies fusion (5,6), recombinant PCR (7), and gene transfection (8). The properties of immunoglobulin isotypes are summarized in *Table 1*.

3. Immunoaffinity separation

The technique of immunoaffinity adsorption relies entirely on the strength and selectivity of binding between an antibody and its corresponding antigen. Generally, antiserum-specific (antibody–antigen) interactions have dissociation constants (K_d) in the region of 10^{-3}–10^{-14} M at 25°C (9). Immunosorbent separations require antibodies that have binding affinities for their antigens (in terms of K_d) of the order of 10^{-10}–10^{-6} M, this value being slightly reduced on immobilization. Antibodies with lower K_d values (high affinities) require harsh conditions to elute antigens from an immunosorbent, which can result in diminished recovery of active product and reduced immunosorbent longevity. Conversely, antibodies with higher K_d values (low affinities) when immobilized would be poor at removing antigens from the liquid phase, resulting in inferior fractionation of solutes and the loss of antigen in the breakthrough from immunoaffinity chromatography (IAC) columns. Low affinity antibodies may be exploited for weak affinity chromatography (10) where a target antigen is particularly labile (see Chapter 3).

The use of an immobilized antibody as a bioselective adsorbent was first reported by Campbell and co-workers (11) over 40 years ago. Techniques employing immobilized antibodies to purify many biologicals (e.g. enzymes, hormones, vaccines, interferons, and antibodies) on a laboratory scale have since been developed and widely reported (12,13). Immobilized antibodies have found particular application in the recovery of biologicals where more conventional methods have failed. The increased availability of mAbs by the development of tools such as the hollow fibre bioreactor, has made immunoaffinity separation a method worthy of consideration for purifying high value products, such as interferon (12), factor VIII (13), and lymphokines (14).

The use of immunosorbents, however, as with all antibody-based processes, does have its drawbacks:

- antibodies have a high molecular weight (e.g. 150000 for IgG), which may limit adsorbate accessibility, while rendering immunosorbents susceptible to denaturation and/or fouling

Table 1. Typical physico-chemical properties of mammalian immunoglobulins

Immuno-globulin	Mol. wt	Serum concentration		Heavy chain	Light chain	Mol. wt of heavy chain	No. of bivalent 'Y' units	Sedimen-tation coefficient	Carbo-hydrate content (%)	E_{280nm}	p*I*	Electro-phoretic mobility
		g/l	μmol/l									
IgG_1	1.46×10^5	8–17	53–113	γ_1	κ, λ	5×10^4	1	7S	2–3	13.8	5.0–9.5	α_2–γ_3
IgG_2	1.46×10^5	3	20	γ_1	κ, λ	5×10^4	1	7S	2–3	13.8	5.0–8.5	α_2–γ_3
IgG_{2a}	1.46×10^5	1.9	13	γ_1	κ, λ	5×10^4	1	7S	2–3	13.8	6.5–7.5	α_2–γ_3
IgG_{2b}	1.50×10^5	1.1	7.3	γ_1	κ, λ	5×10^4	1	7S	2–3	13.8	5.5–7.0	α_2–γ_3
IgG_3	1.70×10^5	1	6	γ_1	κ, λ	6×10^4	1	7S	2–3	13.8	8.2–9.0	α_2–γ_3
IgG_4	1.46×10^5	0.5	2.6	γ_1	κ, λ	5×10^4	1	7S	2–3	13.8	5.0–6.0	α_2–γ_3
IgM	9.00×10^5	0.05–2	0.06–2	μ	κ, λ	6.8×10^4	5	19S	12	12.5	4.5–7.8	β_2–γ_3
IgA	1.70×10^5	1–4	7–27	α	κ, λ	8×10^4	1, 2, or 3	7S	7–11	13.4	4.0–7.0	α_2–γ_3
IgA_1	1.60×10^5	3	18	α_1	κ, λ	5.6×10^4	1, 2, or 3	7S	7–11	13.4	5.2–6.6	α_2–γ_3
IgA_2	1.60×10^5	0.5	3	α_2	κ, λ	5.2×10^4	1, 2, or 3	7S	7–11	17.0	5.2–6.6	α_2–γ_3
IgA_s	3.70×10^5	0.05	0.13	$\alpha_{1,2}$	κ, λ	$5.2–5.6 \times 10^4$	2	11S	11	15.3	4.7–6.2	α_2–γ_3
IgD	$1.70 - 1.84 \times 10^5$	$3–200 \times 10^{-3}$	0.02–1.2	δ	κ, λ	6.8×10^4	1	7S	9–14	14.5	4.7–6.1	γ_1
IgE	1.90×10^5	$3–400 \times 10^{-6}$	$5–2000 \times 10^{-6}$	ϵ	κ, λ	7.2×10^4	1	8S	12	14.0	4.4–7.7	γ_1

- spatial constraints on matrix surfaces may lead to antigen-binding sites (*paratopes*) being sterically hindered
- antibodies specific for an antigen are often not readily available and may need to be produced specifically for a particular immunoaffinity application
- if an immunoaffinity process is to be incorporated into the standard operating procedure for the production of a vital isolate, then, in order to manufacture the required immunosorbent reproducibly, an immortalized cell line producing a fully characterized mAb is essential.

Of the immunoaffinity techniques to be found in the literature, most tend to employ an immobilized antibody component; this will be the emphasis in this chapter.

3.1 Preparation of antibodies for immobilization

A fully optimized immunoaffinity protein separation process is a balance between the resolution of target antigen from its contaminants and the strength of antigen binding required. The quality (purity and specific activity) of an isolated antigen of biological origin will reflect the quality of the immobilized antibody (its purity and condition on immobilization), not its concentration. Therefore, the homogeneity of an immunosorbent requires uniform non-denaturing antibody immobilization and antigen purification processes. Highly substituted immunosorbents (with high concentrations of immobilized antibody) are not commonly produced as these will have high antigen-binding capacities, at the cost of reduced antigen clean-up efficiency due to steric hindrance of paratopes, and elevated non-specific adsorptions (15). Therefore, antibodies are commonly coupled to a matrix at levels well below full saturation in order to achieve maximal capacity with minimal co-recovery of non-specific contaminants.

Polyclonal antibodies (pAbs) contain a number of immunoglobulin sub-clones with different binding affinities and antigen-specificities; this heterogeneity may lead to broad IAC elution peaks under mild elution conditions. Hence, in order to elute a given antigen in a concentrated form from a pAb-derived immunosorbent, strong antibody–antigen dissociating conditions are often required, possibly resulting in damage to either or both components. However, those immunoglobulins most suited to immunoaffinity separation applications (requiring mild dissociating conditions) may be isolated from an antiserum by affinity selection (16). Affinity selection may be achieved by eluting pAbs from an antigen adsorbent by applying a stepwise gradient of decreasing pH (e.g. 0–100% (v/v) 0.1 M citric acid to 0.2 M Na_2HPO_4) or by increasing concentrations of chaotropic agents (e.g. NaSCN, 0–3 M) and selecting the antibody fractions emerging at the weakest dissociating conditions for subsequent immunosorbent synthesis.

When constructing a mAb-secreting cell line, it is also possible to select those clonal colonies producing mAbs which are capable of specific antibody–

antigen interactions whilst, being able to release antigens under mild conditions. Hence, the problems which may be encountered when using pAbs in immunoaffinity processes, such as inappropriate binding affinity and specificity, may be avoided by the careful selection of a mAb. Furthermore, if a mAb immunosorbent is to be used for purifying human therapeutics, potential adverse effects, due to ligand leakage, may be minimized by using a human mAb. Genuine human mAbs are made by transforming cells with Epstein–Barr virus followed by stabilization via repeated recloning. An alternative route is by 'humanizing' a conventional murine mAb-secreting cell line in order to incorporate a human Fc region into the mAbs produced.

When a source of antibodies suitable for immunosorbent synthesis has been identified, an appropriate method of preparing immunoglobulins prior to immobilization is to remove plasma proteins and concentrate antibodies by ammonium sulfate or caprylic acid precipitation. Ammonium sulfate at 40–60% saturation will precipitate antibodies from solution, allowing concentrated antibodies to be recovered by resuspending the pelleted material following centrifugation. Caprylic acid (octanoic acid), 1% (v/v) pH 4, will precipitate most serum proteins but leave immunoglobulin G (IgG) in solution (17). It is advisable to carry out small test-scale (1–5 ml) precipitations prior to performing a precipitation on the bulk of one's sample in order to first optimize the precipitation conditions. Gel filtration allows antibodies to be separated from contaminating plasma proteins on the basis of their relative molecular sizes. Within the Pharmacia range, suitable gels are Sepharose CL-4B (IgG, IgA; resolution range 3×10^4–5×10^5) and Sephacryl S-300 (IgM; resolution range 5×10^4–1×10^6). Anion exchange on diethylaminoethyl- (DEAE-) or quaternary aminoethyl-QAE-conjugated matrices allows immunoglobulins to be resolved from the bulk of contaminating plasma proteins since immunoglobulins will not bind at mildly acid pH, whereas other plasma proteins (particularly albumin) will. Affinity chromatography on immobilized Protein A/G (*Protocol 11*) is possibly the best antibody purification technique prior to immunosorbent synthesis. The final step in antibody preparation prior to immobilization is dialysis against or dilution in an appropriate coupling buffer (18).

3.2 Matrices for immunosorbent synthesis

The selection of a matrix material for immobilizing antibodies will depend largely on the final immunosorbent application desired. A summary of those matrix materials commonly used in the preparation of immunosorbents is given in *Table 2*. Cross-linked beaded agaroses are the most commonly employed supports for immunosorbents because of their macroporosity, low non-specific adsorptions, good chemical stability, relatively low cost, and availability of functional groups for easy immunoglobulin coupling. The drawbacks of these matrices are that they may biodegrade on storage, and their flow rate tends to get slower over a number of cycles due to compaction

Table 2. Immunoaffinity chromatographic matrices

Matrix	Source	Advantages	Disadvantages	Examples
Dextran (α-1,6-linked glucose)	*Leuconostoc mesenteroides* cultures	high ligand capacity on CNBr activation (63), chemically stable; 250 mM NaOH or 20 mM HCl, heat-resistant; to 110 °C (64)	gels are soft, biodegradable volume proportional to ionic strength	Sephadex (Pharmacia)
Agarose (poly-1,3-linked β-D-galactose and 3-anhydro-galactose	marine algae	easy CNBr activation, cheap	biodegradable, chemically degradable, non-specific binding of proteins	Sepharose (Pharmacia) Bio-gel A (Bio-Rad)
Cellulose (1,6-linked, β-1,4-glucose chains)	fibrous plants	solvent-resistant, cheap	high non-specific adsorptions, low permeability; thermal and mechanical stability	Excellulose (Pierce) MemSep (Millipore)
Polyacrylamide (acrylamide copolymerized by *N,N'*-methylene-bisacrylamide)	synthetic	pH stable (1–10) low biodegradation, good mechanical stability	gels adhere to glass (65), low porosity	Bio-Gel P (Bio-Rad) Sephacryl (Pharmacia)
Alumina (particulate Al_2O_3)	synthetic	alkaline-resistant, defined pore sizes, rigid, high flow rates	expensive, difficult to derivatize	Kimal (Anatrace) Macrosorb (Sterling Organics)
Polymer	synthetic	alkaline-resistant, defined pore sizes, rigid, high flow rates	expensive	Macro-Prep (Bio-Rad)
Silica ($SiO_2 \cdot xH_2O$)	mineral extracts	rigid, good mechanical and pore stability, high flow rates	irregular bead shape, hydrolysis at pH > 9, weak cationic exchanger and difficult chemistries	Matrex (Amicon)
Glass (borosilicates)	synthetic	rigid, good mechanical and pore stability, specific pore size and shape, very high flow rate	hydrolysis at pH > 9, difficult chemistries, and weak cationic exchanger	Controlled pore glass, (CPG Inc.); HyperD™ (Upstate Biotech, Inc.)

and/or fouling. Polyacrylmide matrices are superior to those based on agarose, in terms of physical stability and resistance to pressure, allowing their use in FPLC and HPLC. Silica- and glass-based matrices have excellent mechanical stability but poor chemical stability in alkaline conditions. Polymer-based matrices such as Affi-Prep (Bio-Rad) offer an alternative to silica without the poor stability in alkaline conditions. Cellulose-based matrices are not generally used in the production of immunosorbents as they are prone to high non-specific adsorptions, have low permeability and low mechanical and thermal stability.

All of the matrices listed in *Table 2* will produce immunosorbents acceptable for use at laboratory scale. Process-scale IAC purification processes demand a high flow rate (for large volumes of liquid). Quality control of bulk manufacturing processes means that maintenance of an immunosorbent's interparticle porosity is an important factor affecting column longevity. Compaction and/or fouling on application of large volumes of media to an immunosorbent, as in process-scale applications, may reduce the interparticle porosity and hence the permeability to flow. This suggests that rigid materials, such as the synthetic, composite, and polymer preparations, are most likely to be suited to immunosorbents used at process scale.

3.3 Immobilization of antibodies on matrix materials

Some of the more commonly used reagents for coupling antibodies to matrices include:

- 1,4-butanediol diglycidoxy ether (bisoxirane), a homobifunctional reagent which reacts with free amino and hydroxyl groups
- carbonyldiimidazole (CDI), which reacts with free amino and carboxyl groups on a matrix to produce imidazoyl carbamate groups which can react with free amines on antibodies to form stable *N*-alkylcarbamate bonds
- cyanogen bromide (CNBr), which reacts with diol groups to form a stable imido carbonate group which will react with primary amine groups on antibodies to form iso-urea bonds
- divinyl sulfone (DVS), which reacts with the hydroxyl groups on the matrix and binds antibodies via free amino or hydroxyl groups
- tresyl chloride (2,2,2-trifluoroethane sulfonyl chloride, TC), which reacts with hydroxyl groups on the matrix to form a sulfonyl ester, which will subsequently react with amine groups on antibodies with displacement of sulfonate

Traditionally, immunosorbents have been, and still are, produced by CNBr activation. CNBr activation is applicable to all polysaccharide (agarose, cellulose, and dextran) and polyacrylamide matrices, antibodies being attached via their primary amine groups. Matrix activation with CNBr is simple and results in a high level of antibody substitution. The main disadvantage of this reaction is that CNBr is toxic and all reactions involving its use must therefore be carried out in a well vented fume cupboard by trained personnel. Immunosorbents based on CNBr-activated Sepharose may become unstable below pH 5 or above pH 10. These preparations may also contain residual cationic charges, which could lead to the non-specific binding and co-recovery of contaminants in immunoaffinity-purified products (19). This effect may be reduced by including nucleophilic compounds such as glycine in the adsorption and washing buffers, though their presence may also result in an increase in background antibody leakage from agarose preparations (20).

Peng and co-workers (21) found that the most stable binding for immobilizing antibodies to an agarose matrix was via amine or amide bonds, as opposed to the more commonly used CNBr method which results in bonds that slowly degrade. Similarly, a survey of commercially prepared affinity solid phases found that the level of contaminating ligand in an affinity-purified product of biological origin was lowest for those solid phases employing aldehyde/Schiff's base ligand coupling (22).

Ubrich and co-workers (23) compared the stability of immunosorbents prepared by a number of immobilization chemistries. From their findings, five common activation chemistries used for preparing immunosorbents could be ranked from best to worst, on the basis of the amount of antibody bound, DVS $>$ TC $>$ CNBr $>$ CDI $>$ epoxide. In terms of antibody leakage from Sepharose-based immunosorbents, activation methods could be ranked from best (lowest) to worst (highest) in the order TC $<$ DVS $<$ CNBr $<$ CDI. Due to the low capacity of epoxide-activated immunosorbents, Ubrich and co-workers omitted these from their leakage experiments. However, in studies conducted using porous silica (24), it has been shown that epoxide coupling results in immunosorbents with extremely low levels of antibody leakage when compared with those made employing alternative binding chemistries.

Activating agents such as CNBr and triazine are highly toxic, and matrices should be washed thoroughly to remove residual reagents after coupling is complete.

Protocol 1. Cyanogen bromide (CNBr) activation

Warning: This technique may result in production of cyanide gas. In addition to the normal laboratory safety precautions (gloves and coat), the use of a well vented fume cupboard is mandatory.

Equipment and reagents

- Water: reverse osmosis grade or equivalent
- CNBr: 5.0 M solution in acetonitrile (Aldrich)
- Carbonate/bicarbonate (carb/bicarb): 15 mM Na_2CO_3, mM $NaHCO_3$, pH 9.8
- Blocking buffer: 1 M ethanolamine, 0.2 M glycine, pH 7.6
- Screw-capped glass vessels: 50 ml Duran or equivalent
- Sepharose 6B (Pharmacia)

- Antibody solution: 25 mg of immunoglobulins in 25 ml of carb/bicarb, or 25 ml of immunogloulins at 1 mg/ml dialysed against carb-bicarb
- Phosphate-buffered saline (PBS): 20 mM Na_2HPO_4, 137 mM NaCl, 2.7 mM KC1 1.5 mM KH_2PO_4, pH 7.2
- PBS/NaN_3: PBS containing 0.05% (w/v) NaN_3
- Roller mixer: Spiramix 5 (Denley) or equivalent
- Magnetic stirrer

A. *Bead expansion*

1. Weigh 2 g of Sepharose into a 50 ml glass vessel.
2. Add 25 ml of water (reverse osmosis grade) and roller-mix for 16 h at room temperature.

B. *Antibody immobilization*

1. Magnetically stir 25 ml of carb/bicarb in a conical flask on ice for 20 min.
2. Wash 5 ml of expanded Sepharose (see *Protocol 1A*) over a sintered glass filter under suction, by forming a slurry with 25 ml of carb/bicarb five times.
3. Add 5 ml of washed and drained Sepharose to the conical flask and adjust the rate of stirrer-bar rotation in order to mix the beads gently.
4. Mix the solution for 30 min, removing excess liquid and replenishing the ice in the beaker periodically.
5. Add three 10 μl CNBr aliquots to the slurry at 10 sec intervals and stir for 30 min.
6. Pour beads on to a sintered glass filter funnel (still in fume cupboard).
7. Percolate 25 ml carb-bicarb (ice-cold) through beads three times; repeat three times with 25 ml water.[a]
8. Add the drained activated matrix to the antibody solution in a glass vessel and roller-mix for 16 h at 4°C.
9. Pour the beads on to a sintered glass funnel and draw off the liquid under gentle suction.
10. Repeat step 7.
11. Add the beads to 25 ml of blocking buffer in a glass vessel and roller-mix for 2 h at room temperature.
12. Wash coupled gel over a sintered glass filter with 25 ml PBS/NaN_3 five times.
13. Store at 4°C in 25 ml of PBS/NaN_3.

[a] If not immediately required, the activated matrix may now be stored at 4°C for up to a year prior to use.

Protocol 2. Immobilization of immunoglobulins by carbonyldiimidazole (CDI) activation

Equipment and reagents

- CDI (Aldrich) 200 mg in 25 ml of dioxane
- Dioxane stored over anhydrous Na_2SO_4
- Equipment and reagents as in *Protocol 1* (except for CNBr)

Method

1. Wash 5 ml of Sepharose (pre-expanded as described in *Protocol 1A*) over a sintered glass filter under suction by mixing to a slurry successively with 50 ml volumes of water, dioxane–water mixtures (20, 40, 60, and 80%, dioxane (v/v), and dioxane.
2. Add beads to the CDI solution in a glass vessel and roller-mix for 4 h at room temperature.
3. Wash activated beads (as in step 1) with 50 ml volumes of dioxane, dioxane–carb/bicarb mixtures (80, 60, 40, and 20%, dioxane (v/v)', and carb/bicarb.
4. Add activated matrix to 25 ml of antibody solution (as in *Protocol 1*) in a glass vessel and roller-mix for 16 h at 4°C.
5. Pour the beads on to a sintered glass funnel and draw off the liquid under gentle suction.
6. Wash the beads and block unreacted sites as in *Protocol 1B*, steps 9–13.

Protocol 3. Tresyl chloride activation

Equipment and reagents

- Sepharose, expanded as in *Protocol 1A*
- Acetone: anhydrous, stored over anhydrous Na_2SO_4
- Acetone solutions: acetone and water mixed 1:4, 2:3, 3:2 and 4:1 (v/v) respectively to give 20, 40, 60 and 80% (v/v) solutions
- Tresyl chloride (TC; 2,2,2-trifluoroethane sulfonyl chloride): 4% (v/v) TC (99% (v/v), Aldrich) and 2% (v/v) pyridine in dry acetone
- Acetate buffer: 0.01 M sodium acetate, 1 M acetic acid to pH 5.6
- Blocking solution: 1 M glycine pH 7.6
- PBS/NaN_3 (see *Protocol 1*)
- Sintered glass filter
- Screw-capped glass vessel (50 ml)
- Roller mixer

Method

1. Wash 5 ml of expanded Sepharose (*Protocol 1A*) over a sintered glass filter under suction, by mixing to a slurry with 25 ml carb/bicarb five times.
2. Add washed and drained Sepharose to a 50 ml screw-capped glass vessel.
3. Add 25 ml of water and roller-mix for 5 min at room temperature.

4. Allow the matrix beads to form a settled bed and decant off the excess liquid.
5. Repeat steps 3 and 4, replacing water sequentially with 25 ml of the following solutions:
   (a) acetone–water at 20, 40, 60, and 80% (v/v) acetone
   (b) acetone
   (c) TC solution
   (d) acetone–acetate buffer at 80, 60, 40, and 20% (v/v) acetone
   (e) acetate buffer
6. Add antibody solution (as in *Protocol 1*) to the activated beads and roller-mix for 20 h at 18–20 °C.
7. Aspirate off the excess liquid and wash the beads with acetate buffer, as in steps 2 and 3, and once with water.
8. Add 25 ml blocking solution (as in *Protocol 1*) and roller-mix for 16 h at 4 °C.
9. Wash the beads as in step 4, five times with 25 ml PBS/NaN_3 (as in *Protocol 1*).
10. Store at 4 °C in 25 ml PBS/NaN_3

3.4 Commercially available pre-activated matrices

Ideally, immunosorbents should be produced using immobilizing conditions that have minimal effect on the biospecific activity of the antibodies. Suitable immunoglobulin coupling conditions tend to be in aqueous solutions of pH 4–9 (at a pH between 0.5 and 2 pH units from the antibodies' p*I*), ionic strengths between 0.01 and 0.5 M, at temperatures between 4 and 25 °C and with short (less than 16 h) reaction times. *Table 3* lists some of the commercially available activated matrices of defined characteristics suitable for producing immunosorbents.

3.5 Determination of bound antibody

The easiest way to determine the amount of antibody immobilized on a matrix as a result of an immobilizing process would be to measure the concentration of antibody in the immobilizing solution before and after antibody immobilization, by spectrophotometry (25), by protein assay (Coomassie dye binding), or by radiolabelled antibody studies (24). A good estimate of the matrix antibody loading (S; mg/ml) may be gained by purging a column or washing a batch of immunosorbent with binding buffer, collecting the total breakthrough (or washings) and applying the following calculation:

$$S = \frac{(C_o \times V_o) - (C \times V)}{M} \qquad [1]$$

Table 3. Commercially available activated immunoaffinity chromatographic matrices

Activation	Name and suppler	Group coupled	Antibody		Comments
			Conditions	Result[a]	
Cyanogen bromide	CNBr-activated Sepharose 4B (Sigma no. C9142; Pharmacia), CNBr-activated 4% cross-linked Agarose (Sigma no C9142), CNBr-activated Sepharose 6MB (Sigma no. C9267)	$-NH_2$	pH 8–10, 4°C–room temp., 4–16 h; block: 1 M ethanolamine or 0.2 M glycine pH 8, 16 h at 4°C	good Ab immobilisation matrix cationic, average Ab leakage, average Ag binding	good all-round immuno-sorbent bonding stable at pH 5–10, intolerant to nucleophilic compounds; CNBr is toxic
Carbonyldi-idimazole	1,1′-Carbonyldiimidazole-activated agarose attached to 4% cross-linked agarose (Sigma C1150), Reacti-Gel (6X) Support (Pierce, 20260)	$-NH_2$	pH 9–11, 4°C–room temp., 16 h	below-average Ab binding, Ab leakage, and Ag binding	matrix has a long shelf-life (6 months); CDI low toxicity
Epoxy	Epoxy activated-Sepharose 6B (Sigma no. E6754; Pharmacia), Eupergit C (Röhm Pharma), ImmunoPure epoxy-activated agarose (Pierce, 20242)	$-NH_2$ –OH –SH	pH 8–13, 21–25°C, 20 h; stirring not required	low Ab binding, very low Ab leakage, low Ag capacity	matrix very stable on storage; slow binding process

Hydrazide	Affi-Gel Hz (Bio-Rad, 153-6047), carboxymethylcellulose hydrozide (Sigma no. C7760), polyacrylamide hydrazide (Sigma no. P8885), CarboLink coupling gel (Pierce, 20392	–CHO	pretreat matrix with glutaraldehyde, or oxidize antibody with sodium periodate, pH 5–6, 4°C, 16 h	average Ab binding	matrix stable on storage, requires additional activation
N-hydroxy-succinimide	*N*-hydroxysuccinimide activated 4% beaded agarose (Sigma H3512), HiTrap NHS (Pharmacia)	$-NH_2$	pH 9–11, 4°C–room temp., 16 h; block: 1 h 10 mM acetate pH 5.6 or 1 h 1 M ethanolamine	low matrix shelf-life, low Ag binding	intolerant to aqueous conditions
Triazine	Triazine-activated agarose 4XL (Affinity Chromatography Ltd)	$-NH_2$ –OH –SH	pH 4–5, 4°C, 0.5–2 h	above-average Ab binding, average Ab leakage, average Ag binding	toxic agent, uncharged, stable at pH 2–12
Vinylsulfone	Vinylsulphone-agarose (Sigma no. V2755)	$-NH_2$ –OH –SH	pH 8–10, 4°C, 16 h; block: 1 M ethanolamine or 0.2 M glycine, pH 8, 16 h at 4°C	excellent Ab binding	DVS is toxic if leached from column

[a] Ab, antibody; Ag, antigen.

where C_o and C are respectively the initial and final (breakthrough/washings) concentrations (in mg/ml) of antibody in solution, V_o and V are respectively the initial and final volumes (in ml) of antibody solution on immobilization to the activated matrix, and M is the amount of matrix (in ml) to which antibody was bound.

Alternative approaches to determining the level of antibody bound to an immunosorbent are by direct dye binding, or by difference spectroscopy. Direct dye binding assesses the decrease in optical density (at 465 nm) of the supernatant applied to batches of immunosorbent, in a modification of the Bradford protein assay (26). Difference spectroscopy measures the optical density of immunosorbent samples in a dual beam spectrophotometer, blanking against control activated matrix (280 nm).

3.6 Adsorption

In immunoaffinity extraction processes, immobilized antibodies or antigens (or, more specifically, their binding sites: paratopes and epitopes respectively) provide the ligand covalently bound to an immobile matrix (to form an immunosorbent) and their target epitopes or paratopes the ligate. When antibody and antigen are in close proximity, the antibody's paratopes and the antigen's epitopic foci are attracted and orientated, primarily by electrostatic forces. This then elicits the formation of secondary hydrogen bonds, pulling the molecules closer together and excluding water. Finally, van der Waals forces are initiated to form a stable non-covalent bond, target antigens thus being adsorbed (27).

In order to optimize an immunoaffinity separation process, the nature of the adsorptive binding stage needs to be understood. The most suitable conditions for an adsorption process should initially be evaluated by small-scale batch adsorptions of purified antigen on to known amounts of immunosorbent in a 'clean' system (i.e. one where solutions are free of contaminants). The biospecific interactions of immunoaffinity adsorption are considered to be largely analogous to those of antibody–antigen recognition occurring in free solution, which may be described by a reversible equilibrium (Equation 2).

$$K_a\,(Ag + Ab) \underset{k_2}{\overset{k_1}{\rightleftharpoons}} K_d\,(Ag{:}Ab) \qquad [2]$$

where Ag is an epitope on an antigen Ab is the paratope of an immobilized antibody and $Ag{:}Ab$ is a non-covalent complex formed between and antibody and antigen. k_1 and k_2 are rate constants. The ratio of the constants (k_1/k_2) is equal to the effective dissociation constant (K_d), and (k_2/k_1) is the association constant (K_a), of the immunoaffinity system. The rate of antigen adsorption on to an immunosorbent is described by (28):

$$\frac{\delta q}{\delta t} = k_1 C(q_m - q) - k_2 q \qquad [3]$$

where C is the concentration of antigen in solution, q is the immunosorbent concentration of adsorbed antigen per unit settled bed volume, and q_m is the maximum antigen capacity of the antibody solid phase. The equilibrium position of such an adsorption process is therefore:

$$\frac{\delta q^*}{\delta t} = k_1 C(q_m - q^*) - k_2 q^* = 0 \quad [4]$$

where C^* and q^* are the concentrations of antigen in solution, and adsorbed respectively per unit volume of settled adsorbent at equilibrium. From Equation 4, it can be shown that, at equilibrium, the immunosorbent concentration of the adsorbed antigen will vary with the concentration of the antigen in a manner described by:

$$q^* = \frac{q_m C^*}{k_d + C^*} \quad [5]$$

Equation 5 predicts that antigen adsorption should be non-linear, and of the type first described by Langmuir (29), characterized as being a favourable binding isotherm. Rearrangement of this equation gives:

$$\frac{C^*}{q^*} = \frac{C^* + K_d}{q_m} \quad [6]$$

The relationship expressed in Equation 6 can be exploited for the empirical determination of q_m and K_d values for an immunosorbent as demonstrated in *Protocol 4*.

Protocol 4. Determination of the antigen capacity and affinity constant of an immunosorbent by batch isothermal analysis (30,31)[a]

Equipment and reagents

- Donkey anti-rabbit IgG immobilized on controlled pore glass (CAL/DαR, ClifMar Associates) or equivalent
- PBS: see *Protocol 1*
- Rabbit IgG, 2 mg/ml in PBS
- Acetone: stored over anhydrous Na_2SO_4
- Roller-mixer: as in *Protocol 1*
- New 6 ml screw-capped glass vessels
- UV detector: LKB Ultrospec 4050 (Pharmacia) or equivalent
- Aluminium foil

Method

1. Wash 10 ml CAL/DαR, over a sintered glass filter, with successive 50 ml aliquots of water, acetone–water (20, 40, 60, and 80% (v/v) acetone), and finally acetone.
2. Loosely cover the filter with aluminium foil and continue to draw air through the immunosorbent for 16 h, to remove all traces of acetone.

Protocol 4. *Continued*

3. Weigh aliquots of the dry immunosorbent (25, 50, 100, 200, and 400 mg) in duplicate into glass vessels.
4. Add 5 ml of rabbit IgG to immunosorbent-containing vessels and to two control (empty) vessels.
5. Roller-mix for 3 h at room temperature.
6. Assess the level of rabbit IgG remaining in solution (or in the supernatant, after centrifugation) by measuring optical density at 280 nm.
7. Calculate the amount of protein adsorbed per unit mass of solid phase (q^*) from the equation:

$$q^* = \frac{(2 - \textit{final antibody concentration (mg/ml)}) \times 5}{\textit{immunosorbent mass (mg)}} \quad [7]$$

8. Plot q^* against the concentration of protein at equilibrium (C^*) to obtain an adsorption isotherm, the plateau of which is equivalent to the solid phase rabbit IgG capacity (Q or q_m).
9. Alternatively, conduct the above experiment keeping the immunosorbent mass constant (e.g. at 100 mg) and vary the concentration of antigen in solution. Plotting values of C^*/q^* against C^* will result in a straight-line plot which has the slope $1/q^*$ and intercept $-K_d$ on the C^* axis.

[a] Here, a donkey anti-rabbit IgG immunosorbent is used as a model, but the strategy for other immunosorbents would be similar.

In order to prevent errors when calculating the q_m of an immunosorbent due to antigen dilution in the solvent void volume (V_o), this determination should be carried out using dry solid phases. When the immobilized antibody is particularly labile, the V_o of immunosorbents should be pre-determined and estimations of q_m corrected for V_o.

Due to the overall physicochemical properties of the matrix used, antibody, antigen, and/or components of the feed to be purified, non-specific adsorption may also occur. Most commonly, non-specific adsorption occurs as a result of hydrophobic and ionic forces, which must be minimized in order to optimize an affinity separation process (32). Non-specific adsorption may be minimized by:

- avoiding highly substituted immunosorbents
- limiting the length of spacer arms to no more than six carbon atoms (33)
- limiting IAC column size (and therefore capacity) to the amount of antigen to be purified

- conducting adsorptions in semi-dissociating conditions (weak eluting conditions, section 3.7.2; inhibiting all but specific antibody–antigen interactions)
- selective buffering of feed and washing solutions (pH and ionic strength).

Immunoaffinity separations may be conducted using immunosorbents in either column or batch modes of operation. Immunoaffinity chromatography separations result in highly efficient extractions, sometimes at the cost of yield. Such separations tend, therefore, to be used in the production of analytical samples, or therapeutics, or where other methods fail to give sufficient resolution of trace contaminants from the target molecule. Batch immunoaffinity separations maximize the recovery of target solute, but tend to be less efficient at sample clean-up than IAC methods. Batch techniques tend to be used for preparative separations, as a stage in a general purification protocol, or when absolute purity ($<90\%$) is not essential.

3.6.1 Batch techniques

The protocols outlined here are applicable to batch binding antigens to immunosorbents in semi-filled columns or glass vessels. The advantage of using columns is that binding may be achieved in a batch mode while elution and washing stages may be conducted using the solid phase as a packed bed. Consequently, IAC-purified solutes will emerge as concentrated peaks.

Protocol 5. Batch (stirred-tank) mode of adsorption

Equipment and reagents

- 100 ml glass vessels: Duran or equivalent
- All other equipment/reagents as in *Protocol 4*
- Cit/phos: 0.2 M Na_2HPO_4 and 0.1 M citric acid solutions mixed to give solutions of pH 1.8–7.8

Method

1. In a small-scale column affinity purification, establish the correct elution pH (Cit/phos, as in *Protocol 7*) to recover adsorbed antigen from the immunosorbent.
2. Establish the concentration of antigen in the solution to be processed and calculate the amount of immunosorbent required to bind all antigen present (see *Protocol 4*). If the volume of immunosorbent required is greater than half the volume of the antigen solution to be purified, then dilute the antigen solution with PBS (or pre-determined washing/adsorption buffer), such that the volume of antigen is at least twice the immunosorbent volume.
3. Wash the required amount of immunosorbent over a sintered glass funnel by mixing to a slurry with 10 vols of 0.1 M citric acid, then PBS (or appropriate adsorption/washing buffer).

Protocol 5. *Continued*

4. Add the drained immunosorbent to the antigen solution in a glass vessel (or stirred culture vessel).
5. Mix the immunosorbent and antigen solution repeatedly to a slurry, by roller-mixing or stirring with an overhead mixer, for 30 min.
6. Allow the immunosorbent to sediment by gravity and remove the supernatant (collect and keep this as it may require reprocessing).
7. Add two immunosorbent bed volumes of PBS (or chosen washing buffer) and again mix to a slurry for 30 min, as in step 4.
8. Repeat steps 6 and 7 three times using two immunosorbent volumes of pre-determined eluting buffer *but* collect and pool the supernatants containing the affinity-purified antigen at each removal stage. Add 0.1 vols of 10× concentrated PBS buffer to the pooled eluates and dialyse the pool against 100 volumes of PBS/NaN_3.
9. Repeat steps 5 and 6 using, sequentially, three immunosorbent volumes of 0.1 M citric acid and then PBS/NaN_3.
10. Store solid phase at 4°C in PBS/NaN_3.

Protocol 6. Batch adsorption in a column[a]

Equipment and reagents

- Column: 100 ml polypropylene 737-1093 (Bio-Rad) or equivalent
- Immunosorbent: 30 ml immobilized donkey anti-rabbit IgG (e.g. CAL/IAC/DαR, Clif-Mar Associates) or equivalent. The volume used should be in the region of 10–30% (v/v) of the volume of the column chosen
- Antigen: 100 ml, or sufficient to fill the column chosen
- 0.2 M Na_2HPO_4, 0.1 M citric acid, Na_2HPO_4/MeOH, PBS, and PBS/NaN_3 solutions as in *Protocol 5*
- Roller-mixer: Denley Spiramix 5 or equivalent

A. *Column preparation*

1. Prepare column according to *Protocol 7A*, except that in steps 5 and 6 add buffer to the column capacity and roller-mix the gel as a slurry for 5 min at room temperature prior to allowing the buffer to drain through the settled bed under gravity.

B. *Antigen purification*

1. Cap the bottom column outlet and apply antigen solution to the column (pH of antigen adjusted to 7 or dialysed against PBS). Where antigen is dilute and in volumes greater than that of the column (as, e.g., in cell culture supernatant), the antigen may be adsorbed by adding the prepared immunosorbent to clarified antigen solution in a larger glass vessel.

2. Replace the top cap of the column (or seal the larger vessel) and suspend the immunosorbent by repeatedly inverting and agitating gently to form a slurry.
3. Roller-mix the column (or vessel) for 30 min at room temperature.
4. Replace column in the clamp (or stand) and allow the column to form a settled bed.
5. Remove the top cap, *then* remove the bottom cap and allow the antigen solution to drain from the column under gravity. For a larger volume adsorbed in a glass vessel, an aliquot of the slurry can be applied to the column allowing the media to drain from the column outlet to leave the immunosorbent as a settled bed. Add 5 column volumes of PBS to the column and discard the breakthrough.
7. Add sequentially 150 ml (5 column volumes) of Na_2HPO_4 solution, and Cit/phos solutions at pH values of 7.8, 6.7, 4.9, 3.3, and 1.8, collecting each eluate separately.
8. Collect all the eluted fractions and add 15 ml (0.1 vols) of 10× concentrated PBS, dialyse for 16 h (overnight) at 4°C against PBS/NaN_3, then analyse dialysates for antigen of interest.

C. *Column regeneration and storage*

This should be carried out as in *Protocol 7C* applying 150 ml (5 column volumes) of buffer at each stage.

[a]By performing batch mode operations in a column, it is possible to gain the advantages of both complete saturation of all the specific antigen-binding sites and those of antigen recovery in a concentrated form.

3.6.2 Column techniques

The model of immunoaffinity separations presented above for batch binding processes (Equations 2–6) has been considered as too simplistic by many researchers to explain the performance of a process-scale, packed-bed system. More recently, theoretical models, proposed by workers in the area of process design, have attempted to describe packed-bed adsorption and elution stages, taking into account the following adsorbent mass transfer interactions:

- film mass transfer: the transport of adsorbate from the bulk liquid to the stationary phases
- intra-particle diffusion: the transport of adsorbate within the solid phase pores
- binding mass transfer: the ligand–adsorbate interaction

Hence, the overall mass flow during adsorbate and ligand interactions as a solution flows through an immunosorbent may be expressed as follows:

$$\frac{\delta C_B}{\delta t} + \frac{u}{\varepsilon}\frac{\delta C_B}{\delta z} - D_z\frac{\delta C_B^2}{\delta z^2} = \left(\frac{1-\varepsilon}{\varepsilon}\right)\left(\frac{1-\alpha}{R_p}\right) K_L (C_p - C_B) \quad [8]$$

where C_B is the bulk solute concentration (g/ml), t is time (s), u is the superficial velocity (cm/s), ε is the column void fraction, D_z is the axial dispersion coefficient (cm^2/s), z is distance along the column (cm), α is the shape factor (sphere = 2), R_p is the solid phase particle radius (cm), K_L is the fluid film mass transfer coefficient (cm/s) and C_p is the pore solute concentration (g/ml).

The performance of the adsorption stage of a packed-bed IAC process employing immobilized antibodies can best be determined by *frontal analysis*, in which a fixed concentration of antigen (C_o) is continuously applied to the immunosorbent bed inlet, whilst the variation of outlet antigen concentration (δC) is monitored. The variation with time of antigen concentration of the bed outlet is referred to as the 'antigen *breakthrough*', and may be plotted as a *breakthrough curve* (C/C_o against time or volume; see ref. 28). Due to the dynamic nature of IAC adsorption, under these conditions q^* will only be equal to Q when $C_o \gg K_d$. Hence, the q_m and preferred C_o for optimal operation of a packed bed at process scale can be assumed from small-scale batch binding (as in *Protocol 4*), and is given by:

$$q^* = \frac{q_m C_o}{K_d + C_o} = q_m \text{ when } C_o \gg K_d \quad [9]$$

In practice, as C approaches C_o, application of adsorbate is usually stopped (at $C/C_o \approx 0.8$) so as to avoid losing too much antigen to the breakthrough. However, this process does have the disadvantage of leaving a number of free immunoglobulins available for non-specific adsorptions and as a result the co-recovery of contaminants may occur. By evaluating those factors which affect the shape and position of the breakthrough curve of an adsorption process, the most appropriate process conditions for a feedstock may be ascertained.

IAC separations on a small scale (0.2–2 ml bed sizes) may be carried out in disposable columns such as those supplied by Pierce, Bio-Rad, or Lab m; these are supplied with tightly fitting caps and a solid-phase-supporting frit. Columns of this type can be capped and stored at 4°C between uses. An alternative to commercially supplied columns would be to use syringe barrels fitted with a disc of thick filter paper as a frit (34).

By using a stepped Cit/phos gradient elution, it is possible to avoid exposing eluted antigens to lower pH values than are necessary. Employing the above technique, it is possible to recover 6–7 mg of rabbit IgG from 5 ml of 10% (v/v) rabbit antiserum.

Protocol 7. Pulse mode of column operation[a]

Equipment and reagents

- Column: 5 ml disposable (8 mm diameter; no. D823, Lab m) or equivalent
- Immunosorbent: CAL/DαR (as in *Protocol 4* or equivalent
- Citric acid, 0.1 M
- Dialysis tubing: Sigma, D-9277 (10 mm) or equivalent
- Rabbit serum 10% v/v in PBS
- Cit/phos: as in *Protocol 5*, pH 1.8–7.8
- Phos/MeOH: 0.2 M Na_2HPO_4 containing 10% (v/v) methanol
- 10× PBS, PBS, and PBS/NaN_3 as in *Protocol 1*

A. *Column preparation*

1. Allow the purification buffers and immunosorbent to stabilize at room temperature.
2. Clamp the column in a vertical position, cap the bottom outlet of the column, and pipette in 1 ml water, marking the 1 ml level.
3. Suspend immunosorbent beads (in their preserving buffer or in 3 vols of PBS/NaN_3) by repeatedly drawing into and expelling from a 5 ml pipette and add 2 ml of the slurry to the column.
4. Remove the bottom cap and allow the beads to pack by gravity, adding more bead suspension dropwise, to gain a settled bed volume of 1 ml. Do not allow the column to drain dry (alternatively, allow a prepacked column to come to room temperature, remove the top cap, *then* the bottom cap and allow the preserving buffer to drain away under gravity).
5. Add 5 ml of 0.1 M citric acid to the IAC column (adding it slowly down the side of the column so as not to disturb the surface of the immunosorbent) and allow to percolate through under gravity.
6. Re-equilibrate column by adding 5 ml of PBS and allow to percolate through under gravity.

B. *Antigen purification*

1. Apply 10% (v/v) rabbit serum diluted in PBS (or antigen solution adjusted to pH 7 or dialysed against PBS) and allow to percolate through the column under gravity.
2. Apply 5 ml of PBS to the column and discard the breakthrough.
3. Add successively 5 ml Na_2HPO_4 solution and Cit/phos pH solutions at pH 7.8, pH 6.7, pH 4.9, pH 3.3, and pH 1.8 and collect the respective breakthroughs as separate pools.
4. Add (immediately on collection) 0.5 ml of 10× PBS to all the eluted fractions and dialyse for 16 h (overnight) at 4°C against 5 litres of PBS/NaN_3, and analyse dialysates for antigen of interest.

Protocol 7. *Continued*

C. *Column regeneration and storage*

1. Add 5 ml citric acid, Phos/MeOH, and PBS consecutively (the column is now ready for a further purification cycle, go to *Protocol 7B*, step 1, if required).
2. Add 5 ml of PBS/NaN_3 and discard the breakthrough; add a further 5 ml of PBS/NaN_3, replace the caps on the column, and store at 4°C.

[a] Most laboratory IAC purifications are conducted in pulse mode, where feed, washing buffer, and eluting buffer are applied to a column consecutively in pulses.

Protocol 8. Frontal mode of IAC column operation[a]

Equipment and reagents

- Column: 0.1–2 ml glass column (Omni-fit; 7 mm diameter or equivalent)
- Immunosorbent: e.g. 1 ml CAL/DαR (*Protocol 4*) or equivalent
- Solutions: as in *Protocol 7*
- Gradient mixer (model GM1), fraction collector (model Frac 100), and peristaltic pump (Pharmacia) or equivalent
- UV detector: LKB Ultrospec 4050 (Pharmacia) or equivalent

Method

1. Apply 5 column volumes of 0.1 M citric acid (or required eluting buffer) to the column at 0.5 ml (or 0.5 column volumes) per minute. This removes any non-specifically bound proteins resulting from previous immunosorbent uses, microbial activity on storage, and enzymatic degradation of ligand.
2. Change to PBS (or appropriate adsorption/washing buffer) at the pump inlet and add 5 column volumes of PBS to the column at 0.1 ml/min, to allow for renaturation of the immobilized antibodies and re-equilibration.
3. Switch from PBS at the pump inlet to antigen solution (at concentration C_0) and apply antigen to the column at 0.3 ml/min.
4. Collect 1 ml (column volume) fractions from the column outlet and monitor the breakthrough concentration of antigen (C) by OD (at 280 nm), biological activity, or continually monitor C (e.g. using a UV detector fitted with a flow-through cell). Stop applying antigen solution when $C/C_0 = 0.8$. Once measured, the amount of antigen solution required to achieve $C/C_0 = 0.8$ may be applied in future purifications without the need to monitor the breakthrough.
5. Switch solution at the pump inlet to 0.2 M Na_2HPO_4 and apply 5–20 column volumes to wash the column (save these washings along with the breakthrough as both may need to be reprocessed).

6. Apply a Cit/phos gradient (0–100% v/v citric acid) to the column and monitor the breakthrough (as in step 4) for antigen content. Collect the antigen peak, add 0.1 vols of 10× PBS and dialyse this for 16 h (overnight) at 4 °C against 100 vols of PBS/NaN_3.
7. Add > 5 column volumes of citric acid, Phos/MeOH, and then PBS at 0.5 ml/min (the column is now ready for a further purification cycle, (step 3) if required).
8. Add > 10 column volumes of PBS/NaN_3, stop the peristaltic pump, seal the column inlet and outlet, and store at 4 °C.

[a] Frontal analysis is best carried out using a peristaltic pump to apply and control the flow rate (0.1–0.5 column vols/min) of antigen, washing and eluting solutions through an IAC column. Flow rate should be kept low (<0.3 column volumes/min) for the adsorption stage but may be increased to that sustainable by the immunosorbent for washing, desorption, preparation, and regeneration stages. Columns should be packed to their capacity so avoiding excessive solvent mixing in the column reservoir and immunosorbent void volume. All buffers should be made fresh and degassed under a gentle vacuum for 1 h prior to use.

For laboratory-scale IAC separations, a good estimation of Q can be gained by performing a frontal mode antigen purification (as in *Protocol 8*) and calculating the column dynamic (antigen applied (in ml) × C_o, when C/C_o = 0.5) and working (total antigen eluted) capacities.

3.7 Desorption of adsorbates

Desorption (or elution) should ideally result in the rapid recovery of concentrated antigen (within three column volumes); however, this is frequently unachievable with immunoaffinity systems, due to the high binding affinities involved. Eluents must negate the bonding forces which bind the target solute (antigen or antibody) to the immunosorbent. Promoted complex dissociations (increasing k_2) are therefore the most effective means by which to recover antigens, as opposed to inhibition of the forward binding reaction k_1) alone. As a result, elution processes may be classified into biospecific methods (k_1 inhibition) and non-specific methods (k_2 promotion).

3.7.1 Biospecific elution

In the process of biospecific elution, antigen–antibody complexes are exposed to a high concentration of either free competing antibody or non-antigen cross-reactant, the adsorbed antigen being released under the pressure of competition. When the competing solute is present in sufficient excess, antigen is released into the mobile phase. The soluble competing solute may then be separated from the purified antigen by virtue of its different physiocochemical parameters. Biospecific elution is generally considered unsuitable for immunoaffinity systems, since elution would necessitate the addition of large quantities of high-cost free antibody or competing cross-reactant, creating

Table 4. Non-specific elution conditions for immunoaffinity chromatography

Technique	Elution conditions	Comments[a]
Altering solvent pH	Acid range (pH 1.5–4): propionic acid (0.01–0.1 M), glycine–HCl buffer (0.01–0.1 M, pH 1.5–3), or citric acid (0.1 M). Alkaline conditions (pH 11–13.5): KCl (0.055 M), NaOH (0.145 M) pH 13, or NaCl (3M), NH_4OH (0.15 M)	Elution by a shift in pH, may involve only slight alterations in ionic strength, thereby avoiding the denaturation of biological Ags as may occur in other applications. Extremes of pH (>3 pH units from an Ab p*I*) may irreversibly denature labile Abs.
Reversible protein denaturation	Urea (<8 M) and guanidine hydrochloride (<6 M)	Reversibly perturbs protein conformation, allowing them to be used as eluents at neutral pH in immunoaffinity protocols. This technique is particularly advantageous when the Ab or Ag involved is unstable at acid pH.
Polarity-reducing agents	Dioxane (<10% v/v), ethylene glycol (<50% v/v) and methanol (<20% v/v), or acetonitrile (20% v/v) and a pH gradient	The presence of non-polar solvents will reduce hydrophobic binding component of the Ag:Ab interaction. However their use may also affect the stability of hydrophobic bonds maintaining an Ab's tertiary configuration resulting in low recoveries of biospecific activity.
Chaotropic agents	Anions from (best to worst): $CCl_3COO^- > SCN^- > CF_3COO^- > ClO_4^- > I^- > ClO_4^- > NO_3^- > Br^- > Cl^- > CH_3COO^- > SO_4^{2-} > PO_4^{2-}$. Cations (best to worst): $NH_4^+ < Rb^+ < K^+ < Na^+ < Cs^+ < Li^+ < Mg^{2+} < Ca^{2+} < Ba^{2+}$. These salts are used at concentrations of 1.5–8 M	These ions have a 'salting-in' effect, increasing the availability of water molecules, decreasing surface tension, and lowering hydrophobic interactions. They will at high concentrations cause some degree of irreversible Ab denaturation.
Temperature manipulation	Raised temperatures (30–45°C, high K_d values)	K_d values of Ab:Ag interactions are increased by two orders of magnitude when the operational temperature is raised from 4 to 43°C (ref. 25). Adsorption is promoted by low temperatures (2–10°C).
Electrophoretic desorption	Semi-dissociating (weak eluting) conditions, low ionic strength	Ags carrying a net charge (dependent upon pH) migrate towards an electrode of opposite polarity. Affinity-purified protein(s) may then emerge from an IAC column in an electrically focused, concentrated band, separate from any trace contaminants (73).

[a] Ab, antibody; Ag, antigen

further downstream processing problems in the dissociation of antigen–antibody complexes.

A technique has been reported for development of antibodies in which the binding of a first antigen to an antibody paratope causes the release of a second antigen from an adjacent paratope on the same bispecific antibody (35). The creation of bispecific antibodies offers enormous potential for the development of immunoaffinity systems in which biospecific elutions of labile antigens may occur under very mild conditions.

3.7.2 Non-specific elution

Non-specific elution of an antigen from an immunosorbent is achieved by adjusting the environmental conditions such that there is a reversible conformational shift in the tertiary structure of either the antibody or the antigen. The resulting alteration in the physicochemical properties of the mobile phase (buffer) perturbs antibody–antigen complexes; desorbed antigen then emerges from a column closely following the solvent front (within about 0.5 column volumes). *Table 4* lists a number of the non-specific desorption buffering conditions used in immunoaffinity purification.

Harsh eluting conditions can irreversibly denature antibodies, so contact times should be minimized by using small elution volumes or high flow rates. In applications involving unstable antigens or antibodies, it is possible that a combination of eluting agents may perform as mild non-specific eluents. The nature of a satisfactory desorption technique will vary from antigen to antigen and from antibody to antibody. Hence, the most appropriate elution conditions for a given target molecule are therefore best determined empirically.

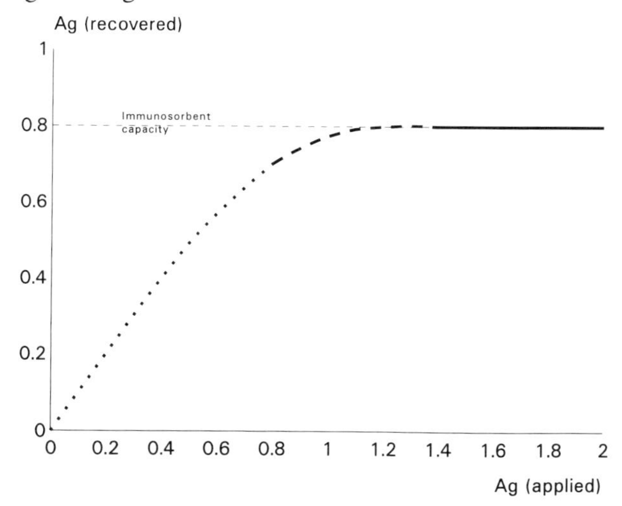

Figure 1. Operating targets for IAC separations: ...., analytical separations, where analyte recovery is more important than absolute purity; - - - - -, general IAC purification performance; —— therapeutic applications, where sample purity is more important than yield.

3.8 Applications of immunoaffinity separations

The design of an immunosorbent used for immunoaffinity separations will depend on the level of antigen recovery and the degree of antigen purity required. *Figure 1* shows the target operational recoveries required for immunoaffinity purifications.

3.8.1 High-performance immunoaffinity chromatography

The use of non-deformable matrices such as those derived from silica and glass as supports when making immunosorbents, allows the principles and tools of high-performance liquid chromatography (HPLC) to be applied to IAC; this has been termed high-performance immunoaffinity chromatography (HPIAC) (34).

Protocol 9. General procedure for HPIAC of protein antigens on immunosorbents (27)

Equipment and reagents

- HPLC system: 0.5 ml immunosorbent packed into glass (Omnifit, as in *Protocol 8*) columns, connected to a Varian 5000 high performance liquid chromatograph, coupled to a Varian UV-50 variable wavelength detector with a Varian 9175 chart recorder, or equivalent equipment
- Phosphate buffer: 0.01 M Na_2HPO_4 pH 7.2
- Antigen sample: 1 ml of antigen solution (0.1–1 mg/ml)
- 0.2 M Na_2HPO_4, 0.1 M citric acid, Na_2HPO_4/MeOH, PBS, and PBS/NaN_3 solutions as in *Protocol 8*

Method

1. Prime HPLC system with phosphate buffer for 10 min at 0.1 ml/min.
2. Switch buffer flow to HPIAC column and detector circuit and prime with phosphate buffer.
3. Set the pump speed to 0.5 ml/min and run consecutively 100 ml of 0.1 M citric acid and 0.2 M Na_2HPO_4 through the system.
4. Inject 100 μl of antigen sample into the injection port and run for 20 min.
5. Adjust the gradient controller to deliver a linear Cit/phos gradient from 0 to 100% 0.1 M citric acid over 5 min (or 10 column volumes).
6. Collect the second peak (and any peaks thereafter; the first peak should be the unbound column breakthrough).
7. Clean the IAC column by applying sequentially 5 ml of 0.1 M citric acid, Na_2HPO_4/MeOH, and 0.2 M Na_2HPO_4. At this point the HPIAC system is ready for applying another sample (step 4).
8. Purge the HPIAC system with PBS/NaN_3, remove the IAC column, seal the ends, and store at 4 °C.

Reproduced with the kind permission of T. M. Phillips (27).

It has been found that the rate-limiting step of HPIAC processes is the rate at which target solute molecules are removed from the mobile phase. All but the adsorption phase of IAC separations may successfully be speeded up in HPIAC processes without compromising the clean-up efficiency of the IAC process.

3.8.2 IAC solid phase extraction (IAC-SPE) of analytical samples

Perhaps the major problem associated with the quantitative assessment of molecules present in biological samples at trace levels (< 1 p.p.m.) is the matrix effects associated with the presence of stoichiometric excesses of other molecules. IAC separations may result in a high degree of analyte recovery from biological samples, typically reducing matrix effects by greater than a thousand-fold in a single purification step (24). Furthermore, IAC extraction allows the analytes present in a large volume of biological fluid (or extract) to be concentrated, thereby effectively reducing the lower limit of detection of an analyte in biological samples. This process is of particular value when the analytes in question are small molecules (e.g., steroid hormones, antibiotics, drugs, and anabolic agents). When incorporated into a routine analytical regime for these residues, IAC systems are relatively simple to operate and automate. An example of IAC is the detection of clenbuterol in tissue samples (*Protocol 10*).

Protocol 10. IAC solid phase extraction (IAC-SPE) of clenbuterol residues from biological samples

Equipment and reagents

- Homogenizer: Silverson or equivalent
- Centrifuge: Beckman J2-21 or equivalent
- 30 ml centrifuge tubes: Sigma (Z14595-5) or equivalent
- Immunosorbent: pre-packed anti-clenbuterol immunosorbent solid phase extraction cartridges (300 μl CAL/IACOSPE/Cb, ClifMar Associates)
- Water bath set at 80 °C
- LP4 tubes: 4 ml disposable polypropylene test tube, Luckhams or equivalent
- SPE vacuum manifold: Baker 10SPE system or equivalent
- PBS and PBS/NaN_3: as in *Protocol 1*
- 10× PBS: 10× concentrated PBS (*Protocol 1*)
- 10% PBS: PBS diluted 1:10 with water
- PBST: PBS containing 0.05% (v/v) Tween 20
- 30 mM HCl
- Biological fluids for testing and control

A. *Sample preparation*

1. Weigh 5 g of finely chopped tissue (muscle, liver, or kidney) into a 30 ml centrifuge tube.
2. Add 15 ml of PBS and homogenize for 1 min.
3. Rinse the homogenizer probe with 5 ml PBS, collect the rinsings, and add them to the tissue homogenate (wash the probe by homogenizing 15 ml of PBS for 1 min three times prior to homogenizing another tissue sample).

Protocol 10. *Continued*

4. Cap centrifuge tube and roller-mix the homogenate for 90 min.
5. Incubate homogenate in a water bath at 80 °C for 15 min, to denature and precipitate the proteins present.
6. Centrifuge for 10 min at 16 000*g* and collect the supernatant (tissue extract).

B. *Column conditioning*

1. Column conditioning is performed, as is sample extraction (below), using 1 ml of clean quality control sample (analyte-free urine or bile) or tissue extract.

C. *Sample extraction*

1. Mount column(s) on a vacuum manifold.
2. Add 2 ml of liquid sample (bile or urine), or 6 ml (2 ml three times) of tissue extract to column and allow to percolate through under gravity.
3. Apply 1 ml of PBST to the column and evacuate under gentle suction.
4. Add 1 ml of 10% PBS to the column and evacuate under gentle suction.
5. Place an LP4 tube under the column.
6. Add 750 μl of 30 mM HCl to the column and complete elution by applying gentle suction and collecting the eluate. Neutralize the sample by adding 250 μl of 10 × PBS. The sample may now be assessed by a quantitative analytical procedure (ELISA, GC-MS, HPLC, or equivalent).
7. Clean column (of hydrophobically bound material) by applying 1 ml of 50% ethanol.
8. Add 1 ml of 30 mM HCl to the column and drain through under suction.
9. Equilibrate and renature the column by applying 3 ml of PBS and allow to drain through under gravity. The column may be re-used (go to section B) or stored at 4 °C in PBS/NaN_3.

Table 5 summarizes the results of clenbuterol extractions using IAC-SPE extraction cartridges for extracting residues of clenbuterol and its related beta-agonists from bovine tissue samples as described in *Protocol 10*. Where residues present are of a metabolized compound, an enzymatic deconjugation step using a mixture of β-glucuronidase and sulfatase (e.g. incubated for 24 h, 37 °C at pH 5.6 with 0.5 U β-glucuronidase from *Helix pomatia*, Sigma) prior to applying sample to a column may be required.

Molecules demonstrating only very slight differences in their surface configuration may be resolved and extracted from biological samples by IAC-

Table 5. IAC extractions of residues of clenbuterol from tissue samples

Tissue	No. of samples	Clenbuterol (p.p.b.)		Yield (%)
		Spiked	Recovered	
Porcine liver	6	0	0	–
	6	0.3	0.209	69.6
	6	0.6	0.621	104
Bovine liver	6	0	0.037	–
	6	0.3	0.210	70
	6	0.6	0.344	57.3
Bovine muscle	6	0	0	–
	6	0.3	0.293	96.6
	6	0.6	0.729	121.0
Bovine bile	105	0	0.056	–

This work was supported by a research contract (IC012) from MAFF.

SPE. *Figure 2* shows the separation of the 17 α- and 17 β-hydroxy isomers of the xenobiotic anabolic agent trenbolone by IAC-SPE. Here trenbolone binding was found to be isotype-specific when a crude tritium-labelled preparation of trenbolone was applied to a column specific for one isomer, indicative of the chiral selectivity achievable when employing IAC-SPE processes.

When packed into solid phase extraction (SPE) cartridges, immunosorbents of the type described above (*Protocol 10*), may be used with the SPE reagents and tools commonly used by analytical chemists to produce extracts for quantitative analysis. It is also possible to increase appreciably the speed of column washing and elution stages, without compromising the overall eluate purity and yield, making it ideal for use as an HPLC-based analytical system. The specificity of IAC would suggest this technique to be applicable to any analyte, limited only by the quality of antibodies produced and their availability.

A drawback to the use of IAC clean-up in quantitative analysis is the effect that residual levels of antibody leaked from IAC columns may have on a quantitative assay. Using 300 μl IAC-SPE cartridges for extracting clenbuterol from bovine bile samples (as in *Table 5*), the level of antibody detected in the eluted fractions occurring as a result of column reuse over 60 cycles was found to be 98 ng/ml (98 ± 17 ng/ml, $n = 60$).

3.8.3 Large-scale IAC and affinity purification of antibodies

The development of a large-scale IAC procedure should ideally begin with the design of an immunosorbent, with the desired end use in mind. The most appropriate form an immunosorbent should take is usually discovered in small-scale laboratory trials (1–5 ml of solid phase), prior to a pilot study with

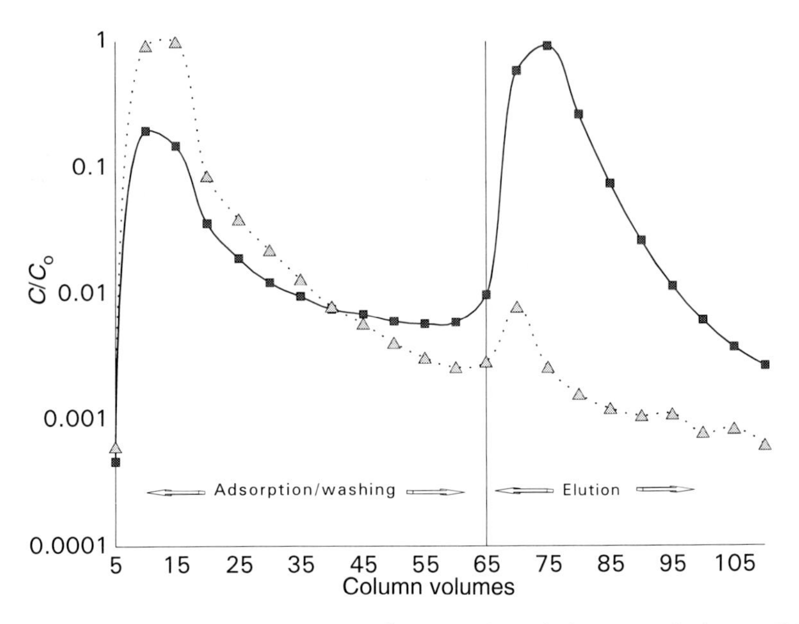

Figure 2. Affinity purification of crude ^{3}H-labelled trenbolone on donkey anti-17α-hydroxytrenbolone (17αOHTb) and on sheep anti-17β-hydroxytrenbolone (17βOHTb). Adsorption/washing: PBS containing 4% methanol; elution: 30 mM HCl containing 20% methanol. Key: C, trenbolone concentration leaving the column; C_o, trenbolone concentration applied to the column; ▲, 17αOHTb IAC-SPE column; ■, 17βOHTb IAC-SPE column.

the same feedstock(s) at preparative scale (10–100 ml of solid phase), and finally scale-up to process scale (1–100 litres of solid phase).

Flow rates may have a considerable effect on the performance of IAC separations, due to the residence time required for the dynamic establishment of antibody–antigen complexes. Excessively high flow rates may result in partial adsorption of antigen, as a result of low contact times. For process-scale purification of a known quantity of a biological, flow rates are usually calculated for minimal column size, taking into account parameters such as C_o and K_a (36). Empirically determined flow rates will need to be lower on prolonged use (or re-use) of columns, due to solid-phase fouling and/or compaction, which will alter the pressure drop over the bed and ligand leakage/inactivation as a consequence of column clean-up. Mass-transfer theory would suggest that the most efficient adsorbents will be those with particle sizes as small as possible. In practice, the choice of process-scale IAC particle size is largely governed by the necessity to maintain high and steady flow rates through often soft and deformable bed materials at low hydrodynamic pressures.

Industrial immunoaffinity purification processes must be cheap, easy, quick, specific, accurate, and precise. Historically, process-scale affinity chromatography has been used as a downstream final clean-up step, selectively removing trace contaminants from commercially prepared biologicals. Ideally, immunoaffinity techniques should be placed further upstream, allowing the bioselective isolation of large quantities of the desired biological, reducing the number of later purification steps needed, and effectively removing the bulk of contaminants in a single operation. This has not been considered a viable proposition in the past, due to the high cost, short life, and very slow flow rates achievable using traditional affinity solid phases. With the development of immunosorbents based on more robust matrices such as Sepharose Fast Flow (Pharmacia), Macro-Prep (Bio-Rad), and Prosep (Bioprocessing) with longer lives, higher flow rates, and lower cost than those previously demonstrated, it is possible to purify biologicals in rapid upstream processes to a high (even therapeutic) grade of purity.

3.8.4 Expanded-bed IAC extraction

The expanded-bed mode of immunosorbent operation is a system of adsorption chromatography that exploits all the advantages of batch and packed-bed modes of adsorbent use (see Chapter 1, Section 3). Expanded-bed IAC extraction is achieved by packing an immunosorbent into a column such that a settled bed occupies no more than 30% of the total column volume. Fluid is then passed upwards through the bed at a flow rate, usually expressed in terms of a superficial velocity (U, in cm/min), equal to the volumetric flow rate (ml/min) divided by cross-sectional area (cm^2). When U is increased, gravity is opposed such that immunosorbent particles become supported by the fluid and the bed expands and becomes fluidized. Under these conditions, a target molecule may be extracted from a solution on to suspended adsorbent beads and the beads washed. Elution may then be achieved by stopping the solvent flow, allowing the column to form a settled bed, reducing the column volume to that of the adsorbent, and applying eluting buffer in normal packed bed mode (*Figure 3*).

In order to optimize this process for purifying a particular antigen-containing feed, the following must be determined:

- the dynamic capacity of the fluidized immunosorbent and hence the amount of solid phase required to purify a particular antibody
- the bed expansion achieved as a function of the superficial velocity (U) of solvent
- the operating U producing fluidization with acceptable antigen recovery and, by taking into account the previous two points, select the appropriate vessel size.

4. Bacterial Fc receptors

The utility of immobilized bacterial Fc receptor adsorbents for the affinity purification of antibodies for therapeutic applications is widely acknowledged

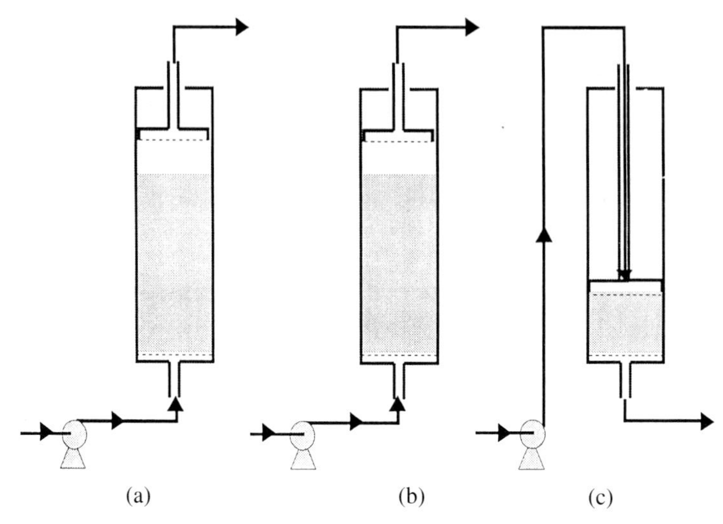

Figure 3. Schematic diagram of expanded-bed operation for protein purification. (a) Expanded-bed adsorption of target molecule; (b) expanded-bed washing of adsorbent with adsorbed target molecule; (c) packed-bed elution of target molecule. (Reproduced from ref. 46 with permission.)

(37). Bacterial Fc receptors are cell-surface proteins and are classified into one of six groups determined by their reactivity with the IgG sub-classes of mammalian species (38).

Protein A (SpA) is a type I Fc receptor found in *Staphylococcus aureus*. Commercially, native-SpA is isolated from cell membrane of cell cultures of *S. aureus*, by lysostaphin hydrolysis of glycosidic bonds in the peptidoglycan layer followed by ion-exchange chromatography clean-up. Recombinant SpA prepared from genetically engineered colonies of *Escherichia coli* strain has also been developed and is commercially available (e.g. from Repligen, USA).

The SpA molecule consists of a single polypeptide chain containing little or no carbohydrate. It has a molecular weight of 42 000, a p*I* of 4.85–5.15, and UV adsorption maximum at 275 nm, $E_{1\%} = 1.65$. It contains four tyrosines and no tryptophan residues, which has been implicated in its irregular mobility on SDS–PAGE. The SpA globular protein has a markedly extended shape and is very resistant to denaturing agents such as 6 M guanidine HCl, 70% ethanol, and 0.1 M HCl.

The SpA molecule has five homologous regions, each of around 50 amino acid residues; four of these (regions A, B, C, and D) contain a single receptor site capable of binding to an IgG molecule with high affinity; the fifth region (E) has low affinity Fc binding and is also the site of attachment to the cell wall. Region E has also been associated with IgG Fab region binding (39). SpA's IgG binding is greatest for human IgG_1, which has a binding affinity constant of 4.8×10^7 litres/mol. Limited binding also occurs with IgM and IgE from some species, recognition occurring solely in the Fc region of the antibody molecule. The nature of the SpA–Fc interaction, as revealed by crystallographic refinement, is due to the formation of a major hydrophobic link between residues of the SpA molecule's active site and the antibody's C_H2 and C_H3 domains in the Fc region. In addition, there is a minor polar link in which C_H3 residues are involved in a sulfate ion link (40).

Protein G (StpG) is a type III Fc receptor of group C streptococci. It has been suggested as an alternative ligand to SpA for the isolation and purification of immunoglobulins, as it has been found to bind a wider range of IgG subclasses than Protein A (including human IgG_3, mouse IgG_1, rat IgG_{2a}, IgG_{2b}, and sheep IgG_1). StpG is isolated from cultures of streptococcal cells by purification of the supernatant or by trypsin digestion of the cell membranes. Protein G has a molecular weight in the region 35–40 kDa. and a UV maximum at 273 nm (37). DNA sequencing of native StpG has identified two locations at which immunoglobulins are bound along with sites for albumin and cell surface binding. Recombinant Protein-G molecules have been constructed where the albumin and cell binding sites have been removed, reducing non-specific binding during immunoglobulin purification (41). Historically, due to the relative costs of SpA and StpG preparations and their respective use, SpA adsorbents were most commonly reported for affinity purification of antibodies. Additionally, the strong binding of bovine IgG to Protein G suggested that SpA was a better ligand than Protein G for purifying mAbs cultured in the presence of fetal calf serum.

A recombinant fusion protein called Protein A/Protein G is now commercially available (ICN Biomedicals, USA). This protein is derived from a hybrid gene comprised of *S. aureus* Cowan I strain Protein A and the binding domains of *Streprococcus sp.* Lancefield group G Protein G strain. While more expensive than the more commonly used SpA, recombinant and synthetic ligands are reputed to offer a wider range of immunoglobulin sub-class specificity, and increased adsorbent capacity while operating in a manner similar to both immobilized SpA and StpG preparations. *Table 6* summarizes the affinity of SpA and StpG for immunoglobulins from different species.

Protein H (StpH) is a type II Fc receptor protein found in group A streptococci, which may be expressed in recombinant *E. coli* (42). StpH consists of a single polypeptide chain of molecular weight 42–45 kDa. It selectively binds human IgG and rabbit immunoglobulins, and does not bind mouse, rat, cow, sheep, goat, or human IgA, IgD, IgE, or IgM. For this reason it has been identified as a ligand ideally suited to the purification of human and humanized antibodies.

4.1 Preparation of Fc receptor adsorbents

Immobilization chemistries used in the commercial SpA solid phases include cyanogen bromide (CNBr), triazine, aldehyde/Schiff's base and a number of

Table 6. The Affinity of Protein A and Protein G for immunoglobulins from different species[a]

Species	**Subclass**	**Affinity of SpA**[b]	**Affinity of StpG**[b]
Bovine	IgG_1	weak	*strong*
	IgG_2	strong	strong
Chicken	IgG	none	none
Dog	IgG	*strong*	weak
Goat	IgG_1	weak	*strong*
	IgG_2	strong	strong
	IgM	none	none
Guinea pig	IgG_1	strong	strong
	IgG_2	strong	strong
Horse	IgG	weak	*strong*
Human	IgG_1	strong	strong
	IgG_2	strong	strong
	IgG_3	weak	*strong*
	IgG_4	strong	strong
	IgA_2	*weak*	none
	IgM	*weak*	none
Mouse	IgG_1	weak	weak
	IgG_{2a}	strong	strong
	IgG_{2b}	strong	strong
	IgG_3	strong	strong
	IgM	none	none
Pig	IgG	*strong*	weak
Rabbit	IgG	strong	strong
	IgM	none	none
Rat	IgG_1	weak	weak
	IgG_{2a}	none	*strong*
	IgG_{2b}	none	*weak*
	IgG_{2c}	strong	strong
	IgM	none	none
Sheep	IgG_1	weak	*strong*
	IgG_2	strong	strong
	IgM	none	none

[a] Differences in binding are shown in italics. Data collected from references 24, 27, and 49.
[b] SpA, *Staphylococcus* Protein A; StpG, *Streptococcus* Protein G.

undisclosed chemistries. *Table 7* summarizes some of the commercially available SpA solid phases. Protocols for producing immobilized Fc receptor adsorbents in the laboratory are the same as those used for immunosorbent synthesis (see Section 3.3). Füglistaller (43) found, on comparing commercially available SpA adsorbents, that ligand leakage was lowest for those

Table 7. The composition and performance of some commercially available *Staphyloccus* Protein A (SpA) solid phases

Solid phase	Matrix composition	Coupling method	SpA	Linear flow (mm/min)	Capacity mAb (mg)[a]	SpA/mAb ($w/w \times 10^{-6}$)
ACL[b]	4% cross-linked agarose	triazine	'unknown'	8.4	5.3	195
Bio-Rad	'polymer'	'unknown'	native	2.9	3.5	25
NYGene	cellulose	'unknown'	recombinant	13.7	5.1	21
Pharmacia Fast Flow	4% cross-linked agarose	cyanogen bromide	native	8.4	5.0	15
Prosep-A HC	controlled pore class	aldehyde/Schiff base	native	9.9	15.0	12
Repligen	6% cross-liked agarose	aldehyde/Schiff base	recombinant	7.0	2.2	12
Sepharose CL-4B	4% cross-linked agarose	cyanogen bromide	native	6.5	4.3	20
Sigma Fast Flow	4% cross-linked agarose	cyanogen bromide	native	10.8	5.6	16

[a] Average capacity of duplicate 1 ml columns (7.3 mm diameter) and a NYGene cartridge over four cycles with a low saline washing/adsorption buffer (glycine 1 M, sodium chloride 150 mM; pH 8.6). [b] Trial sample.

employing alkylamine or ether immobilizing linkages. In a similar SpA leakage study (22), the level of contaminating SpA found in affinity-purified mAbs on frontal analysis was lowest when using those commercially available SpA adsorbents employing aldehyde/Schiff's base ligand bonding.

It is generally accepted that recombinant receptor proteins such as recombinant Protein A have lower binding affinities than their native receptor proteins, therefore adsorbents should have higher non-specific adsorptions due to the presence (at antibody capacity) of non-specific interaction sites. However, studies conducted in our laboratories have shown no significant difference between the use of native or recombinant Protein A when assessing the IgG binding of SpA adsorbents prepared in-house, or when assessing commercially available SpA adsorbents (22,24).

Figure 4 shows the effect SpA substitution had on the capacity of mAb (murine IgG_I) and SpA contamination of purified mAb for SpA adsorbents based on porous silica (SpA-PS) (24). As may be seen those SpA-PS preparations with moderate mAb capacities ($Q \approx 8$ mg/ml) purified mAbs which had lower levels of residual SpA contamination than those with either much higher or lower capacities. This finding would suggest that, when used to purify antibodies destined for therapeutic applications, large SpA-PS columns of average capacity used at high flow rates may prove more appropriate than using smaller, highly substituted adsorbents.

The purification of immunoglobulins on bacterial Fc receptor adsorbents

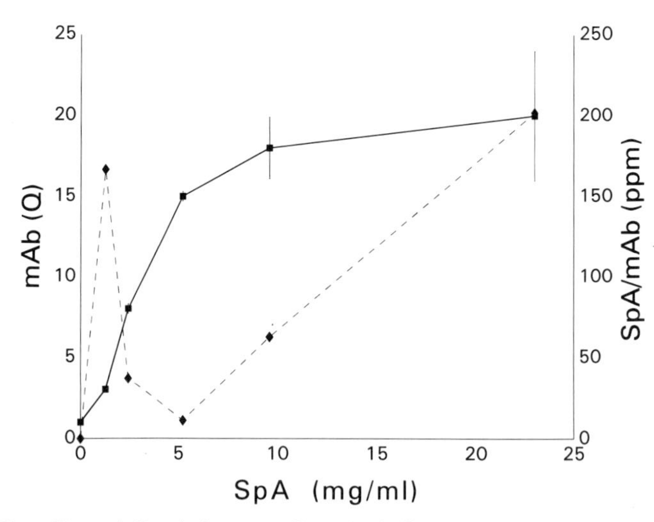

Figure 4. The effect of *Staphylococcus* Protein A (SpA) substitution level on the mAb capacity and leached SpA levels in mAbs affinity-purified on immobilized SpA adsorbents. Adsorbents were assessed at mAb saturation (frontal analysis) for mAb capacity (squares, solid line: *Q*, mg/ml) and the SpA (diamonds, dashed line: SpA/mAb, p.p.m.) content of purified mAbs ($n = 4$).

(SpA and StpG) demands efficiency of an affinity system in which both the ligand and ligate are macromolecules (35–45 and 150–900 kDa respectively). In such a system, maintaining the intra-particle porosity of the matrix is an important factor affecting adsorbent longevity. Compaction and/or fouling on application of large volumes of media may reduce the intra-particle porosity and hence the empirically practicable flow. This suggests that rigid matrices are most likely to be suitable for process-scale applications. Another approach to the problem of solid-phase compaction is to use columns with a large diameter:height ratio.

4.2 Purification of immunoglobulins using immobilized Fc receptors

The degree to which an antibody-containing sample is pre-treated prior to purification on an Fc receptor adsorbent largely depends upon the nature of the starting material (e.g. ascites, tissue culture supernatant, serum, or plasma). Additionally, having only small volumes of the antibody-containing solution available may prohibit extensive pre-treatment prior to applying the solution to an Fc receptor column.

Protocol 11. Laboratory-scale purification of a mAb (murine IgG_1) on immobilized SpA and StpG adsorbents[a]

Equipment and reagents

- Column: 8 mm diameter (Lab m: no. D823 or equivalent)
- Adsorbent: Prosep-A 'High Capacity' (Bioprocessing, Protein G-Sepharose 4 Fast Flow (Pharmacia) or equivalents
- Cit/phos and Phos/MeOH: see *Protocol 7*.
- PBS, PBS/NaN_3, and 10× PBS: see *Protocol 1*
- Adsorption and washing buffer: for SpA columns (22), 1 M glycine, 0.15 M sodium chloride, pH 8.6; for StpG columns (44), 0.05 M Tris–HCl, 3 M NaCl, pH 7.8
- mAbs-containing cell supernatant

Method

1. Same as for *Protocol 7A*.
2. Add 5 ml of 0.1 M citric acid to the column (draining it slowly down the side of the column so as not to disturb the surface of the adsorbent) and allow to percolate through under gravity.
3. Re-equilibrate the column by adding 5 ml of adsorption buffer and allow to percolate through under gravity.
4. Apply mAb solution and allow to percolate through the column under gravity.
5. Apply 5 ml of washing buffer to the column and discard the breakthrough.
6. Add 5 ml of Na_2HPO_4 solution, Cit/phos solutions at 1:3, 1:1, and 3:1 (citrate:phosphate ratios, v/v), and citric acid solution.

Protocol 11. *Continued*

7. Collect all the eluted fractions, add 0.5 ml of 10× PBS and dialyse for 16 h (overnight) at 4 °C against PBS/NaN_3; analyse dialysates for mAb of interest.
8. Add 5 ml aliquots of citric acid, Phos/MeOH, and PBS (the column is now ready for a further purification cycle (step 9) if required).
9. Add 5 ml of PBS/NaN_3 and discard the breakthrough.
10. Add 5 ml of PBS/NaN_3 containing methanol (5% v/v), replace the caps on the column, and store at 4 °C.

[a] The procedure described here is for small-scale mAb purifications carried out in disposable columns (as described in Section 3.8.1). This procedure is equally applicable to automated HPLC and FPLC applications.

High concentrations of inorganic salt (NaCl > 1 M) promote the binding of immunoglobulin sub-classes which otherwise bind weakly to SpA (45). Additionally, the mAb capacity of SpA solid phases is maximized when using buffers containing a high level of inorganic salt. However, on affinity purifying mAbs, using adsorption and washing buffer conditions of high salt and pH 8–9, the levels of contaminating SpA, albumin, and transferrin in the purified mAbs are also elevated, due to their promoted non-specific hydrophobic bindings (24). The efficiency of mAb purification operations using immobilized SpA adsorbents may be increased by decreasing the inorganic salt content of the adsorption and washing buffers (< 0.2 M) and introducing glycine (up to 2 M).

In expanded-bed experiments (see Section 3.8.4) conducted using Prosep-A 'High Capacity', the dynamic capacity of the solid phase for a humanized monoclonal IgG_1 antibody has been found to be >80% of the packed-bed dynamic capacity irrespective of superficial velocity (46). In theory, it should be possible, by a series of valves, to process cell culture broths containing whole cells by a refinement of this method. This technique offers great potential for the development of aseptic on-line purification techniques for processing antibody-containing cell culture supernatants.

4.3 Ligand leakage from SpA adsorbents

The potential physiological effects of injecting contaminating levels of SpA (47) present in SpA adsorbent affinity-purified mAbs means that SpA leakage is of concern to those producing therapeutic mAb preparations. On frontal analysis using SpA adsorbents, SpA leached from adsorbents is prevented from being washed away with free mAb in the mobile phase by the relatively higher concentration of mAb adsorbed on the stationary phase (22). This results in a fall in the level of SpA leaving the column. On approaching column capacity ($q = Q$), the level of SpA leakage appears to

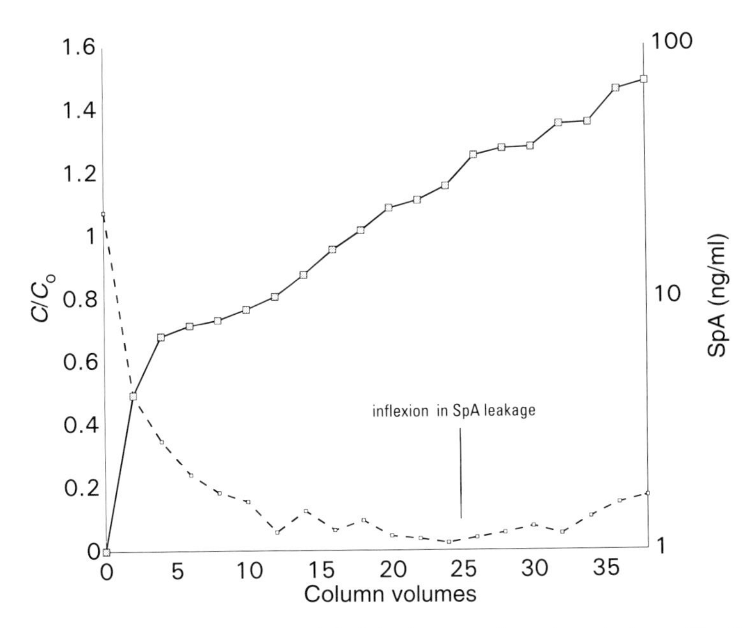

Figure 5. Antibody breakthrough and SpA leakage on frontal analysis purification of a mAb using Prosep A 'High Capacity' 1 ml column. *C*, mAb concentration leaving the column; C_0, mAb concentration applied to the column; large squares, solid line: C/C_0; small squares, dashed line: SpA (ng/ml).

increase again slowly to the original rate. This would suggest a dynamic replacement of immobilized SpA-associated mAb with free mAb. The leaking SpA accumulates on the binding sites of the immobilized mAb, emerging as a peak of leakage on elution of the antibody. An optimal elution point would therefore be at the point of inflexion in the SpA leakage curve on frontal analysis, which approximates to $C/C_0 = 0.8$ (*Figure 5*).

Figure 6 shows the level of SpA contamination of a mAb purified on Prosep-A 'High Capacity'. Over eight cycles, this process resulted in SpA contamination levels which fell from an initial level of $> 140 \times 10^{-}6$ to a maintained lower level of $< 7 \times 10^{-6}$ (SpA/mAb, w/w) by the fourth purification cycle. This observed leakage-stabilizing effect was consistent with the finding of Francis and co-workers (48) for Protein A–Sepharose. Preconditioning SpA affinity solid phases by a pilot small-volume mAb purification prior to a process-scale purification of mAbs from tissue culture supernatant may prove the best regime for the purification of therapeutics.

4.4 High-performance liquid affinity chromatography (HPLAC) of antibodies on immobilized SpA

Chicken serum antibodies and egg yolk IgYs do not normally interact with SpA or mammalian Fc receptors (16), suggesting that the chicken is an ideal

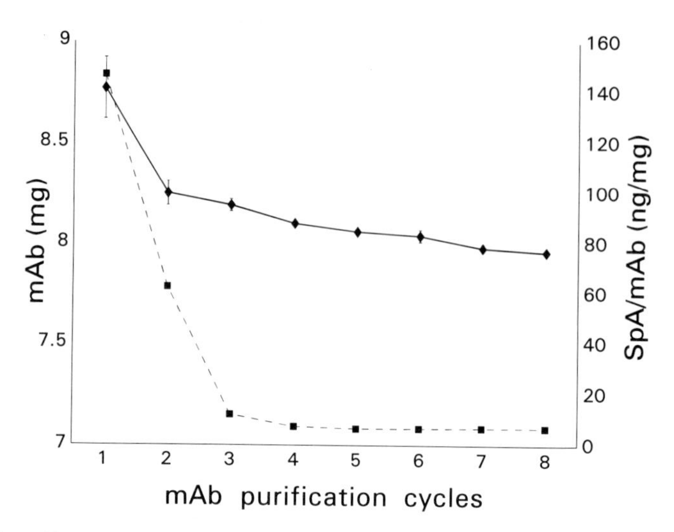

Figure 6. mAb recovery and SpA contamination on purifying 5 ml of tissue culture supernatant over eight purification cycles using a 1 ml column of Prosep A 'High Capacity'. mAb recovered (diamonds, mg) and the SpA/mAb (squares, ng/mg) content of purified mAbs (n = 4).

species in which to raise a specific antiserum to bacterial Fc receptors. *Figure 7* represents the HPLAC of chicken egg yolk extract containing anti-SpA antibodies (IgY) on Prosep-A 'High Capacity'. Anti-SpA IgYs, eluted from the SpA adsorbent by a gradient of decreasing pH, typically produce four elution peaks. This elution profile appears consistent with that found by Ohlson and Niss (36) who used this technique to isolate populations of circulating immune complexes from the blood of patients with immune-complex associated diseases. HPLAC on immobilized SpA has also been used as a means of fractionating pAbs into their various immunoglobulin isotypes. When the eluted peaks of IgY from an HPLAC fractionation were assessed by IEF, all appeared to be free of contaminants and of the same p*I*, indicative of their being of the same isotype, and separated on the basis of binding affinity to SpA (*Figure 7d*).

The column residence time of an antibody-containing solution required for antibody adsorption to take place is rate limiting. Hence, an increase in the recovery of antibody by HPLAC at constant superficial velocity (U) and volumetric flow rate, is achieved by increasing the column length. Increasing U, while purifying antibodies from serum on a Prosep-A 'High Capacity' column, results in a proportional increase in the breakthrough component of the purification profile. All other stages of the affinity separation could successfully be speeded up to flow rates of >30 column volumes/min with little effect on the overall efficiency of the isolation.

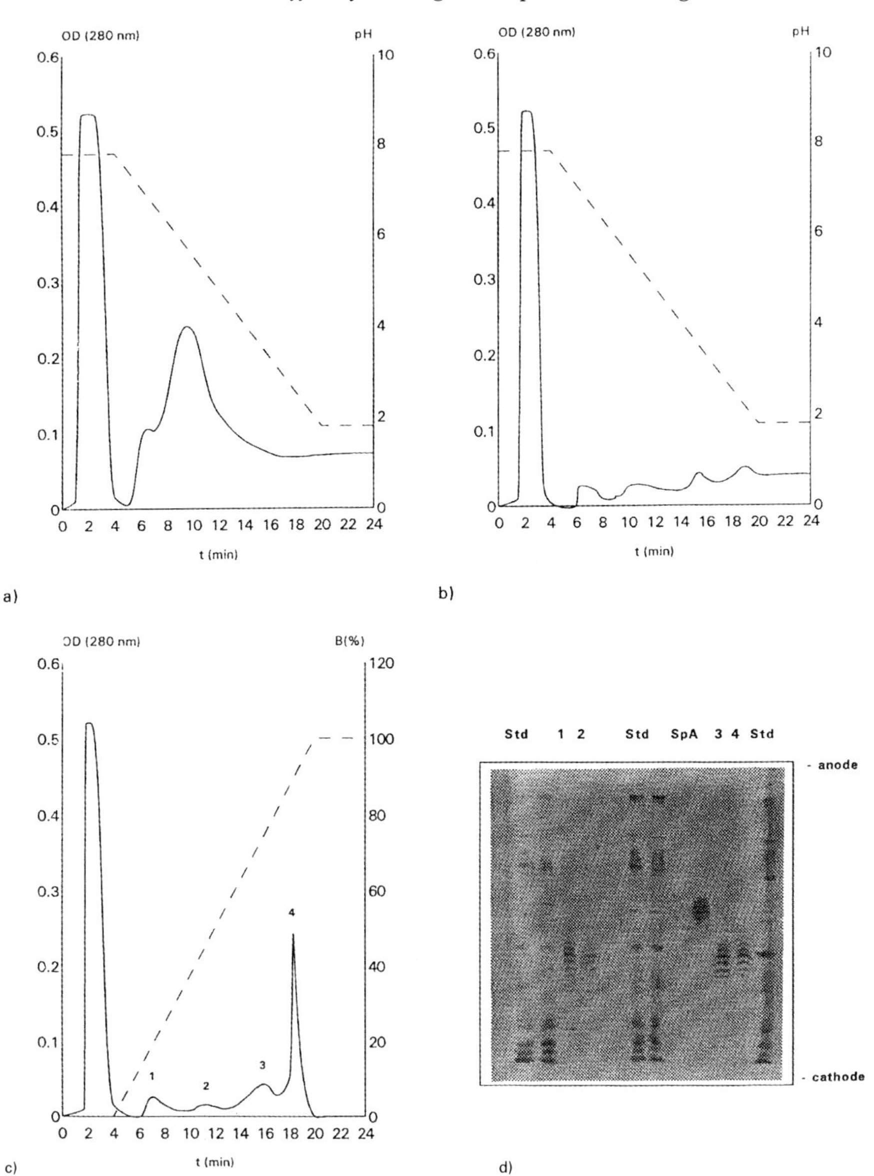

Figure 7. HPLAC of crude egg yolk extract containing chicken anti-Protein A (CαSpA) antibodies on a 0.5 ml Prosep A 'High Capacity' column. Two hundred microlitres of CαSpA were injected at a flow rate of (a) 0.5 ml/min (– – – –, pH; ——, OD) and (b) 2 ml/min (– – – –, pH; ——, OD). Eluent: linear gradient from buffer A: (200 mM Na_2HPO_4, pH 7.8) to buffer B (100 mM citric acid pH 1.8) (c) Gradient as in (b) except that buffer B was 1 M glycine–HCl, pH 1.6. Optical detection: 280 nm (– – – –, B%; ——, OD). Fractions around the protein peaks of run (c) were collected and assessed by iso-electric focusing electrophoresis (Pharmacia: PhastGel 3-9, Separation Technique no. 100, silver stain kit) shown in (d). Std = isoelectric point standards.

Protocol 12. General procedure for the HPLAC of antibodies on immobilized SpA[a]

Equipment and reagents

- High pressure column: glass, adjustable volume, 0–3 ml (Omnifit)
- Buffers/solutions: 0.2 M Na_2HPO_4, 0.1 M citric acid, adsorption and washing buffer (22), Cit/phos, Na_2HPO_4/MeOH, PBS and PBS/NaN_3; prepared as in *Protocol 7*.
- HPLC adsorbent: Prosep-A 'High Capacity' (Bioprocessing) or equivalent
- HPLC system: HPLC, variable wavelength detector, and chart recorder: Varian models 500, UV-50, and 9175 or equivalent.
- mAb-containing sample solution

Method

1. Prime HPLC system with 0.1 M citric acid.
2. Connect adsorbent column in-line between the pump and the detector.
3. Set the pump speed to 0.5 ml/min and flush the system for 20 min or until 10 column volumes of solvent have passed through the column.
4. Switch the solvent inlet to adsorption buffer and flush the system for an equal length of time as in step 3.
5. Inject 200 μl of sample into the injection port, switch the injection loop position to on-line with the system, and run for 10 min.
6. Apply a linear gradient from 0 to 100% Cit/phos elution buffer over 5 min, or half the flushing time used in step 3.
7. Discard the breakthrough peak and collect the second peak.

[a]The method detailed in this protocol is for a 0.5 ml column of adsorbent; the amount of solvent will require adjusting for different column sizes.

When using Fc receptor adsorbents in an HPLAC mode, a problem which has often been reported as occurring is one of elution peak 'ghosting'. The occurrence of these peaks may be reduced by diluting (or dialysing) the sample in adsorption buffer, and ensuring the system buffers are fully degassed. Using the materials and conditions suggested in *Protocol 12*, ghost peaks were eliminated by applying $>$ 10 column volumes of 0.1 M citric acid over a period of 30 min. Unexpectedly, the elimination of ghost elution peaks here proved to be a dependent on the 'time of clean-up' and not on 'clean-up solution volume' (24).

4.5 SpA column regeneration and clean-up

For process-scale applications, adsorbent stability (chemical and mechanical) may be the most significant factor in determining the overall cost of a separation method, followed closely by specificity and binding capacity (19). The initial outlay for a solid phase must be weighed against those factors affecting

Table 8. The effect of sanitizing procedures on the *Staphylococcus* Protein A (SpA) and monoclonal antibody (mAb) content of Prosep-A 'High Capacity' purified mAb (afp27 sz)

Buffer	Protein eluted		SpA/mAb (w/w × 10^{-6})
	SpA (ng)	mAb (mg)	
80°C dry (72 h)	13.5	3.6	3.75
Acetone	94.2	13.6	6.93
$CHCl_3/CH_3OH/H_2O$ (5:4:1 by vol.)	101.6	12.9	7.88
Methanol	101.8	11.5	8.85
Low saline (ref. 14)	111.9	12.2	9.17
Acetic acid (10%)	120.4	11.8	10.20
GuHCl (6 M)	114.2	10.1	11.31
Urea (6 M)	133.4	11.6	11.50
NaOH (1 M)	143.2	12.2	11.74
Autoclaving	89.7	3.7	24.24

20 ml of concentrated mAb (708 μg/ml) containing tissue culture supernatant added to columns after roller-mixing column(s) in 5 vols of the above buffers, or heat sterilizing and re-equilibrating. Ranked in order of increasing SpA contamination of mAbs (24).

adsorbent longevity, its compressibility, non-specific adsorption, ligand leakage, and matrix stability and, hence the matrix cost per batch separation. Resistance to 1 M sodium hydroxide (used in depyrogenation) has been considered a prerequisite of affinity media for use in this area, excluding controlled pore glass preparations (50). *Table 8* lists the effects of sanitation on the performance of two immobilized SpA-PS adsorbents. By withstanding heat sterilization processes (autoclaving, or 80°C dry for 72 h), these adsorbents may allow the aseptic purification of antibodies destined for therapeutic applications. However, this may also result in a reduced purification efficiency due to the increased presence of denatured SpA, which lowers adsorbent capacity, increases ligand leakage, and provides sites for non-specific protein interactions.

4.6 Purification of antibodies after Fc receptor affinity adsorption

After affinity SpA adsorbent purification, antibodies will invariably be >95% pure. Labile antibodies should rapidly be adjusted to neutral pH by the addition of a concentrated buffer such as 1.5 M PBS or 1 M Tris base. Eluted antibodies should be dialysed against an appropriate buffer prior to storage. Residual contaminants present (often at below detectable levels) may include non-immunoglobulin proteins, ligand leached from the adsorbent, and antibodies degraded during the purification process. These may be removed from purified antibodies by affinity processes such as dye-ligand affinity chromatography, or by physicochemical means using gel filtration or ion-exchange chromatography.

For short-term storage (< 1 year) a preservative such as azide (0.01% w/v) may be added to purified antibodies and solutions stored at 4 °C. For long-term storage, antibodies are best aliquoted into small batches (to prevent repeated freeze–thawing damage) and stored frozen at below −20 °C. When freezing antibodies the use of phosphate buffers such as PBS should be avoided as these will decrease in pH as a sample is frozen, risking antibody hydrolysis. Lyophilization of antibodies will ensure stability on storage for many years; however, due to the tendency of mAbs to aggregate during the lyophilization process, this should be avoided where possible.

5. Miscellaneous immunoglobulin-binding ligands

Novel applications of mAbs in therapy, tumour imaging, and immunoaffinity separations, indicate an exponential increase in their demand to a market in excess of $6 billion per annum by the year 2000 (51). Along with the increase in demand for highly purified antibodies has been the development and refinement of a vast array of affinity techniques able to fractionate antibodies from biological fluids, removing contaminants which might otherwise impair their function or therapeutic acceptability. In this section, examples of the more commonly used affinity adsorbents are listed along with suggestions for their use.

5.1 Hybrid and mimetic ligands

Low molecular weight antibody-binding ligands and antibody receptor proteins have been developed to produce adsorbents which are more robust and versatile than bacterial receptor adsorbents. Two such ligands are Avid AL (UniSyn Technologies Inc.) and KappaLock (Upstate Biotechnology Inc.).

Avid AL is a low molecular weight, synthetic, antibody-binding ligand offering all the advantages of a dye-ligand adsorbent (Section 5.2) with the added bonus of being antibody-specific (52). Avid AL can be sanitized with caustic and should not be potentially antigenic to man as are Proteins A, G, and A/G. It selectively binds antibodies in conditions of neutral pH and low salt such as PBS (pH 7.4), or tissue culture supernatant adjusted to neutral pH. Avid AL adsorbents have antibody capacities comparable to those of Proteins A and G, will bind a broader range of immunoglobulin isotypes than either Protein A or G and, unlike Proteins A and G, will also bind chicken IgG and egg yolk IgY.

KappaLock is a 32 kDa recombinant protein (expressed in *E. coli*) that binds specifically to the kappa light chains of antibodies, irrespective of heavy chain isotype. Being a ligand specific for the kappa chain region of antibodies its use is applicable to the purification of antibody fragments ($F(ab')_2$, Fab, and Fv) as well as all the antibody classes. Adsorption and elution conditions are the same as those which may be used for Protein A or G purifications, the antibody capacity being dependent on the abundance of kappa light chains in an antibody sample.

5.2 Dye-ligand affinity chromatography

The use of textile dyes, in the production of adsorbents (see Chapter 3, Section 4), offers many significant advantages over the more conventional immobilized biological ligands, in that:

(a) They are produced on a large scale and are therefore readily available of a standard quality at very low costs.
(b) They are easily coupled directly to matrices via their reactive functions resulting in a fairly stable bond, thereby avoiding hazardous activation chemistries.
(c) They are physically and chemically stable, tolerating the presence of detergents, enzymes, and depyrogenating agents and are unaffected by heat sterilization processes.
(d) The protein-binding capacity of immobilized dye adsorbents is much greater than that of natural biological media (10 to 100-fold), and proteins may be resolved under mild conditions producing a good yield.
(e) Dye columns are readily re-usable.
(f) Being mechanically (and chemically) stable, dye columns may be operated at high flow rates, pressures, and temperatures.

All these properties would suggest dye ligands as the ideal choice of ligand for the process-scale clean-up of therapeutic antibodies.

The main drawback to the use of dye ligands for affinity purifying antibodies is the heterogeneity of dye–protein binding. Consequently, it may be necessary to screen a number of different immobilized dyes, to find the one most suited to isolating antibodies from the feed in question. Another problem reported to be associated with the use of dye-ligands is one of excessive levels of ligand leakage (24,53). In view of the different chemical structures of the various dyes available and their diverse interactions with proteins, it is generally accepted that their sequential use in increasing the target protein content of fractions of a variety of plasma proteins is the best approach to their exploitation. Histocially, dye-ligand affinity adsorbents have not been very effective for binding antibodies. Most protocols employing dye-ligand affinity in antibody purifications do so in order to remove trace contaminants from antibody preparations (e.g. removal of albumin on Cibacron Blue F3G-A adsorbents) at a downstream stage in a purification procedure (see Chapter 3, *Protocol 7*). 'Mimetic' (Affinity Chromatography Limited) ligands have also been derived from synthetic dyes to incorporatc features which mimic the protein-binding components of natural ligands. 'Mimetic' ligands can also be used for negative affinity purification when applied to antibody separation procedures (54).

Protocol 13. Purification of chicken egg yolk IgY by negative affinity purification on Rivanol-PS (55)[a]

Equipment and reagents

- Column: Lab m no. D823 or equivalent
- Adsorbent: 1 ml CAL/PS/Riv (ClifMar Associates) or equivalent
- Adsorption and washing buffer: sodium acetate 10 mM, pH 5.6.
- PBS: 20 mM Na_2HPO_4, 0.137 M NaCl, 2.7 mM KCl. 1.5 mM KH_2PO_4, pH 7.2
- Egg yolk containing IgY
- Eluting buffer: 2% NaCl (w/v) in 0.1 M lactic acid
- 3 M $MgCl_2$ solution
- PBS/NaN_3: PBS containing 0.05% (w/v) NaN_3
- Roller mixer
- 20 mM phosphotungstic acid

A. *Yolk preparation*

1. Dilute egg yolk 10% (v/v) in washing buffer and roller-mix for 2 h.
2. Centrifuge diluted yolk at 2000 *g* for 10 min at 4°C and collect the IgY-containing supernatant.
3. Add 1% (v/v) $MgCl_2$ (3 M) and 5% (v/v) phosphotungstic acid (20 mM) dropwise while stirring.
4. Incubate, standing at 37°C for 90 min, then repeat step 2.
5. Dialyse IgY supernatant against three changes of 100 vols of washing buffer.

A. *Method*

1. Allow affinity column (1 ml Riv-PS adsorbent) and all the purification buffers to come to room temperature.

2–4. The conditioning step and steps 3 and 4 are as in *Protocol 6*.

5. Replace the bottom column outlet cap and apply 5 ml of egg yolk IgY supernatant to the column.
6. Replace the top inlet column cap and roller mix the supernatant and column matrix for 15 min at room temperature.
7. Allow the solid phase to form a settled bed again.
8. Remove the top column cap, then the bottom cap, and collect the IgY-containing breakthrough.
9. Apply 1 ml of washing buffer to the column and collect the breakthrough; repeat twice.
10. Apply 5 ml of eluting buffer to column and discard the non-IgY eluate.

11. Add 5 ml of eluting buffer to the column and discard the breakthrough.
12. Add successively 5 ml each of water/acetone at 2:1, 1:1, and 1:2 (v/v).
13. Add 5 ml washing buffer and discard the breakthrough (the column is now ready for a further antibody clean-up cycle if required).
14. Add 5 ml of PBS containing azide (0.02%, PBS/NaN_3) and discard breakthrough.
15. Add 5 column volumes of PBS/NaN_3, replace the caps on the column, and store at 4 °C.

[a]Using this technique, it is possible to extract egg yolk immunoglobulins (IgYs) by negative affinity purification of crude egg yolk (0.01% v/v). However, this process is limited by the rapid clogging of the adsorbent due to its lipophilic nature. Using this protocol, it has been possible to recover more than 2 mg of >95% pure IgY per ml of egg yolk.

5.3 Thiophilic antibody purification

Thiophilic adsorption chromatography (TAC) is a highly selective lyotropic salt-promoted protein adsorption process which has been extensively investigated and reported by Porath and co-workers (56). The stationary phase of TAC adsorbents are synthetic, hence the physical/chemical stability of adsorbents is dictated by the stability of the matrix material used (see *Table 2*). Thiophilic adsorption occurs as a result of proteins recognizing a sulfone group in close proximity to a thiol ether. Thiophilic adsorption offers a low cost and more selective alternative to ammonium sulfate precipitation, promoting both hydrophilic and hydrophobic interactions. Thiophilic adsorptions are promoted by high concentrations of neutral water structure-forming salts (non-chaotropic ions, *Table 2*) which decrease the availability of water in solution. Therefore, the adsorption process is a salting-out process, promoted by salts such as potassium and ammonium sulfate. In the presence of high concentrations of these salts, thiophilic adsorbents bind IgG, IgM, and IgA and some minor serum components (particularly α_2-macroglobulin), but not serum albumin. Unlike hydrophobic interaction chromatography (HIC), thiophilic adsorption yields better defined separations. Typically, TAC is able to produce an 80% yield of >95% pure immunoglobulin from a crude serum sample (57). However, immunoglobulin adsorption depends on salt concentration: increasing the $(NH_4)_2SO_4$ or K_2SO_4 concentration of serum feed on a TAC column from 0.7 M to 1.5 M will also increase the co-adsorption of non-immunoglobulin proteins (even albumin) to the adsorbent. Additionally, high $(NH_4)_2SO_4$ or K_2SO_4 concentrations will decrease the immunoglobulin content of feed solutions due to their salting-out of solution.

Elution is achieved by gradually adjusting solvent $(NH_4)_2SO_4$ or K_2SO_4 concentration and collecting the immunoglobulin-rich fraction (*Table 10*).

Protocol 14. Thiophilic adsorption purification of antibodies[a]

Equipment and reagents

- Thiophilic adsorbent (T-Gel Pierce & Warriner), Prosep-thiosorb (Bioprocessing Ltd), or equivalent), 2 ml
- General chromatographic equipment as in *Protocol 7*
- Chromatographic column: 6.6 mm diameter, 60 mm height
- Wash buffer: 0.02 M Hepes, 0.5–1 M NaCl, 0.6 M $(NH_4)_2SO_4$, pH 7.5 (adjust with 1 M NaOH)
- First elution buffer: 0.02 M Hepes, 1–2 M NaCl, 0.6 M $(NH_4)_2SO_4$, pH 7.5 (adjust with 1 M NaOH)
- Second elution buffer: 60% ethylene glycol

A. *Antibody preparation*

1. For small volumes (<5 ml) of serum, ascites, or culture supernatant, dialyse against 100 volumes of wash buffer. For larger volumes of serum, ascites, or culture supernatant, add Hepes, NaCl, and solid $(NH_4)_2SO_4$ to attain the wash buffer concentrations.
2. Centrifuge (>5000*g*) or filter (0.2–0.4 μm) the antibody solution.

B. *Antibody purification*

1. Prepare a column of thiophilic adsorbent as in *Protocol 6*, washing with wash buffer.
2. Apply antibody solution (60 cm/h, or under gravity flow).
3. Add 10 ml wash buffer and discard the breakthrough.
4. Add 1 ml of the first elution buffer and discard.
5. Add 6 ml of the first elution buffer and collect the breakthrough.
6. Repeat steps 4 and 5 using the second elution buffer.

[a] Most mAbs will be eluted in the first elution buffer. Polyclonal antibodies will require the second elution stage to ensure that all hydrophobically adsorbed antibodies are recovered.

5.4 Hydroxyapatite

Hydroxyapatite is a crystalline form of calcium phosphate which is able to fractionate plasma proteins on the basis of complex interactions between its Ca^{2+} groups and negatively charged moieties on the surface of proteins (58,59). Beaded ceramic hydroxyapatite (CHT; Bio-Rad) has been developed by the controlled sintering of calcium phosphate crystals at high temperatures. Proteins bind to CHT in solutions of low ionic strength (10 mM phosphate pH 6.8) and are desorbed at high ionic strength (0.5 M phosphate, pH 6.8). Binding affinities are a function of a protein's surface charge density, proteins therefore being eluted in a gradient of phosphate in ascending order of p*I* (60). *Figure 8* shows the laboratory- and preparative-scale isolation of murine IgM from ascites on CHT.

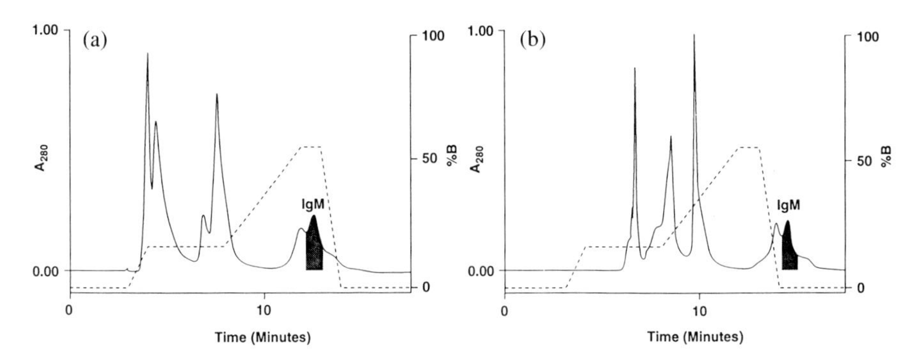

Figure 8. Purification of IgM-containing mouse ascites (diluted 2:1 with 10 mM phosphate, pH 6.8: buffer A) on Macro-Prep Ceramic Hydroxyapatite (CHT Bio-Rad), type I, 10 μm. (a) 0.25 ml of ascites on 2 ml Bio-Scale CHT2-I column; (b) 2.3 ml of ascites on 20 ml Bio-Scale CHT20-I column. Buffer B: 0.5 M phosphate pH 6.8. This figure is reproduced by kind permission of Bio-Rad Laboratories, USA.

5.5 Adsorbents for purifying IgM mAbs

Due to their tendency towards low stability, specificity, and affinity, these antibodies are best avoided for all but specific applications (18). The easiest way to purify a mAb of the IgM class is by precipitation in conditions of low pH (4–6) and ionic strength (1–10 mM). Alternatively, IgMs may be purified by weak anionic exchange, though this requires higher ionic strengths than for IgGs and may result in greater plasma protein co-recovery. Affinity adsorbent techniques most suited to the purification of IgM mAbs include thiophilic adsorption (Section 5.3), hydroxyapatite (Section 5.4), and lectin adsorption.

The lectin mannan-binding protein from mammalian serum is able to bind IgMs via any free mannose or *N*-acetylglucosamine groups (61). The binding of IgM to mannan-binding protein adsorbents requires the presence of Ca^{2+} (20 mM $CaCl_2$ in 1 M NaCl buffered to pH 7.8). Elution of IgM is effected by using the same buffer devoid of Ca^{2+} ions. The lectin GNL from the snowdrop (*Galanthus nivalis*) binds to the α-D-mannosyl terminal of the carbohydrate component of IgM molecules (24). (See Chapter 5 for details of lectin-affinity chromatography.)

5.6 Adsorbents for purifying IgA mAbs

IgA-class mAbs are best purified by DEAE anion-exchange chromatography; this requires higher ionic strengths than for IgGs, but lower than those required for IgMs and may result in minimal plasma protein co-recovery. Alternatively, IgAs may be successfully purified on a lectin adsorbent using immobilized Jacalin (62) (see *Table 9* for conditions).

Table 9. Miscellaneous immunoglobulin-binding ligands

Ligand	Source	Specificity	Ab-purifying conditions		Capacity (mg/ml)	Comment
			Adsorption	Elution		
Jacalin	jackfruit lectin	*O*-linked oligosaccharides; IgA_1 but not IgA_2, fetuin, chorionic gonadotrophin, and plasminogen	T-buffer: 0.5 M NaCl, 0.2 M Tris–HCl, 0.1 M lactose, pH 7.5	T-buffer, replacing lactose with melibiose (66) or 0.5–0.8 M galactose (67,68)	10	Best used to purify IgA. May separate crude IgA into IgA_1 and IgA_2 (68).
T-Gel	synthetic: $-CH_2CH_2SO_2CH_2CH_2X-R$, where X is an atom with an electron lone pair (57)	thiophilic adsorption: IgG, IgM, and IgA	0.02 N Hepes, 0.5–1 M NaCl, 0.6 M K_2SO_4 or $(NH_4)_2SO_4$, pH 7.5	Buffer 1: 0.02 M Hepes, 1–2 M NaCl, pH 7.5. Buffer 2: 60% ethylene glycol	20	Polyclonal Abs require the use of buffer 2 to elute immunoglobulin partially bound by HIC. Conditions on Ab and source.
Protein H	recombinant (*E. coli*) cell wall protein derived from group A *Streptococci*	FcRII receptor for IgG Fc (human IgG_1, IgG_4, and rabbit IgG, but not mouse) (42)	0.05 M Tris–HCl, 1 M glycine, 0.15 M NaCl, pH 7.4	1 M glycine–HCl, pH 2, or 0.1 M citric acid	11	Selective for human IgG and rabbit Abs, will not bind immunoglobulins from mouse, rat, cow, sheep, goat, and other human immunoglobulin classes.
Hydroxy-apatite	synthetic: Ca^{2+}, PO_4^{3-}	selectivity dependent on phosphate concentration.	0.01 M phosphate, pH 6.8	0.5 M phosphate, pH 6.8 and/or pH gradient (5.5 up to stability of protein)	>25	Mild conditions, purity 50–90%, good for IgM isotypes (59,69) e.g. Macro-Prep Ceramic Hydroxyapatite (Bio-Rad).
KappaLock	recombinant (*E. coli*) protein	κ chain specific, four Ab-binding domains	TBS: 0.05 M Tris–HCl, 0.15 M NaCl, 0.05% NaN_3, pH 7.4, 4°C	elution: 0.05 M glycine-HCl, pH 2.7; neutralisation: (elute into 0.1 vols) 1M Tris–HCl, 1.5 M NaCl, 1 mM EDTA, 0.5% NaN_3, pH 8	>20	Binds all Abs depending on their % κ light chains. e.g. Upstate Biotech. Inc., TCS Biomedicals.

Protein A	native cell wall protein from *S. aureus*, recombinant (*E. coli*) protein (RepliGen, USA; Porton Products, UK)	Fc region of Abs, four domains (40,70); Fab region of Abs one domain (39)	low saline: 1 M glycine, 0.15 M NaCl, pH 8.6; high saline: 1.5 M glycine, 3 M NaCl, pH 8.9	Cit/phos pH < 4, 0.1 M citric acid or 0.1 M glycine-HCl, pH 2; neutralization: (elute into 0.2 vols) 1 M Tris base, 1.5 M NaCl, 0.02% Thimerosal, pH 8	10	Well documented, easy to use good chemical stability (*Table 8*). e.g. Protein A-Sepharose 4 Fast Flow (Pharmacia Biosystems) or Prosep-A High Capacity (Bioprocessing Ltd).
Protein G	group G *streptococci*	Fc region of Abs (70)	PBS: 20 mM Na_2HPO_4, 0.137 M NaCl, 2.7 mM KCl, and 1.5 mM K_2HPO_4, pH 7.2	0.1 M citric acid or 0.5 M acetic acid/$(NH_4)OH$ to elution pH; neutralization: (elute into 0.2 vols) 1 M Tris base, 1,5 M NaCl, 0.02% Thimerosal, pH 8	20	Native protein G has an albumin binding site (71), removed in recombinant Protein G, Higher Ab binding affinity than Protein A. Always use a neutralizing buffer. e.g. Protein G Sepharose Fast Flow.
Avid AL	synthetic (73)	K_d:Fc $>$ IgG $>$ Fab	PBS (0.15 M, pH 7.2)	0.1 M acetate, pH 3 or 0.4–0.9 M triethylamine pH 7.4	8	Resistant to extremes of pH and autoclaving and proteases. Binds egg yolk IgY (UniSyn Tec Inc.).

The conditions, capacities, and recoveries quoted are typical for extracting antibodies (Ab) from tissue culture supernatant.

Table 10. Typical capacities of thiophilic adsorbents

Antibody	Capacity (mg/ml)		Purity (%)	
	Prosep[a]	T-Gel[b]	Prosep[a]	T-Gel[b]
Polyclonal antibodies				
Bovine IgG	10	17.9	95	90
Calf IgG		11.1		89
Sheep IgG	15	12.3	96	89
Goat IgG	11	17.9	97	92
Rabbit IgG	8	6.7	74	84
Horse IgG	10	13	89	93
Human IgG	11	8.8	92	70
Pig IgG	12	21.1	97	90
Mouse IgG		8.6		63
Rat IgG		13		79
Chicken IgG		5.2		76
Dog IgG		12.2		91
Guinea Pig IgG		11.1		71
Monoclonal antibodies				
Mouse IgG_1	8	11.6	98	92
Mouse IgG_{2a}		9.3		88
Mouse IgG_{2b}		9.8		97
Mouse IgG_3		10.7		94
Humanized IgG_1	18		98	
Mouse IgM	1.4		95	

[a] Prosep-thiosorb, Bioprocessing Ltd.; [b] T-Gel, Pierce and Warriner (UK) Ltd.
Reproduced from company technical literature by kind permission.

6. Conclusion

Immunoaffinity techniques are rapid, easy, capable of high yield recovery and resolution to >90% purity of target molecules from complex biological mixtures. Immunoaffinity chromatography (IAC) is applicable to process-scale procedures such as the isolation of therapeutic proteins from tissue culture supernatant and smaller batch-scale applications such as the separation of analytical samples prior to quantifying by direct coupling to an analytical column (GC or HPLC) or indirect use of an assay (ELISA, GC-MS). A compromise between the speed of sample processing, the cost, and the degree of reproducibility is needed. When applied to the clean-up of analytical samples, IAC is limited only by the quality of the antibodies available and the performance of immunosorbents when immobilized on a matrix. The additional advantages of using IAC sample clean-up from an analytical point of view is that the bioselective resolution of molecules can exclude a number of molecules that might interfere with an absolute assessment of the analyte content such as isomeric forms, and metabolized analogues. Furthermore, by selecting antibodies with the appropriate cross-reactivities or by using a mixture of specific antibodies, it becomes possible to use immunosorbents for the selective recovery of multi-residues or multi-therapeutics from complex mixtures.

Acknowledgements

I should like to thank Brian Morris and Piotr Kwasowski for their technical advice in the preparation of this chapter.

References

1. Hurn, B. A. L. (1974). *Br. Med. Bull.*, **30**, 26.
2. Catty, D. (1985). *Antibodies I: a practical approach.* IRL Press, Oxford.
3. Catty, D. (1989). *Antibodies II: a practical approach.* IRL Press, Oxford.
4. Land, R. B., Morris, B. A., Baxter, G., Fordyce, M., and Forster, J. (1982). *J. Reprod. Fert.*, **66**, 625.
5. Groves, D. J., Morris, B. A., and Clayton, J. (1987). *Res. Vet. Sci.*, **43**, 253.
6. Groves, D. J., Sauer, M. J. Rayment, P., Foulkes, J. A., and Morris, B. A. (1990). *J. Endocrinol.* **126**, 217.
7. Lewis, A. P. and Crow, J. S. (1991). *Gene*, **101**, 297.
8. Bruynck, A., Seemann, G., and Bosslet, K. (1993). *Br. J. Cancer*, **67**, 436.
9. Capra, J. D. and Edmundson, A. B. (1977). *Sci. Amer.*, **236**, 50.
10. Ohlson, S., Lundbland, A., and Zopf, D. (1988). *Anal. Biochem*, **169**, 204.
11. Campbell, D. H., Luescher, E., and Lerman, L. S. (1951). *Proc. Natl Acad. Sci. USA*, **37**, 575.
12. Mutami, M. and Dyer, M. C. (1986). *J. Biol. Chem*, **261**, 1309.
13. Zimmermann, T. S. and Fulcher, C. A. (1986). US Patent no,. 4361509.
14. Ambrus, J. L., Jr and Fauci, A. S. (1985). *J. Clin. Invest*, **75**, 732.
15. Absolom, D. R. (1981). *Sep. Purif. Methods*, **10**, 239.
16. Godfrey, M. A. J., Kwasowski, P., Clift, R., and Marks, V. (1992). *J. Immunol. Methods*, **149**, 21.
17. Russ, C., Callagaro, I., Lanza, B., and Ferrone, S. (1983). *J. Immunol. Methods*, **65**, 269.
18. Campbell, A. M. (1991). In: *Laboratory techniques in biochemistry and molecular biology* (ed. P. C. van der Vliet), Vol. 23. Elsevier, London.
19. Narayanan, S. R. and Crane, L. J. (1990). *Trends Biotechnol.*, **8**, 12.
20. Dertzbaugh, M. T., Flickinger, M. C., and Lebherz, W. B., III (1985). *J. Immunol. Methods*, **83**, 169.
21. Peng, L., Calton, G. J., and Burnett, J. W. (1986). *Enzyme Microb. Technol.*, **8**, 681.
22. Godfrey, M. A. J., Kwasowski, P., Clift, R., and Marks, V. (1993). *J. Immunol. Methods,* **160**, 97.
23. Ubrich, N., Hubert, P., Regault, V., Dellacherie, E., and Rivat, C. (1992). *J. Chromatogr.*, **584**, 17.
24. Godfrey, M. A. J. (1993). Assessment of silica-based solid phases in bioseparations. PhD Thesis. University of Surrey, UK.
25. Little, J. R. and Donahue, H. (1967). In *Methods in immunology and immuno-*

chemistry (ed. C. A. Williams and M. A. Chase), Vol. 2, p. 343. Academic Press, New York.

26. Bradford, M. M. (1976). *Anal. Biochem.*, **72**, 248.
27. Phillips, T. M. (1989). In *Advances in chromatography* (ed. E. Grushka and P. R. Brown), pp. 133–73. Marcel Dekker, New York.
28. Chase, H. A. (1984). *J. Chromatogr.*, **297**, 179.
29. Langmuir, L. J. (1916). *J. Amer. Chem. Soc.*, **38**, 2221.
30. Roe, S. (1989). In *Protein purification methods: a practical approach* (ed. E. L. V. Harris and S. Angal), pp. 175–244. IRL Press, Oxford.
31. Chase, H. A. (1984). *Chem. Engng Sci.*, **39**, 1099.
32. O'Carra, P., Barry, S., and Griffin, T. (1974). In *Methods in enzymology*, vol. 34 (ed. W. B. Jackoby and M. Wilchek), pp. 108–26. Academic Press, London.
33. Steers, E., Cuatrecasas, P., and Pollard, H. (1971). *J. Biol. Chem.*, **246**, 196.
34. Phillips, T. M., More, N. S., Queen, W. D., and Thompson, A. M. (1985). *J. Chromatogr.*, **205**, 327.
35. Randle, B. (1992). European Patent Application 92303734.5.
36. Hubble, J. (1987). *Biotech. Bioengng*, **30**, 208.
37. Boyle, M. D. P. and Reis, K. J. (1987). *Bio/Technology*, **5**, 697.
38. Palmer, D. A., French, M. T., and Miller, J. N. (1994). *Analyst*, **119**, 2769.
39. Ibrahim, S., Kaartinen, M., Seppälä, A., Matoso-Ferreira, A., and Mäkelä, O. (1993). *Scand. J. Immunol.* **37**, 257.
40. Deisenhofer, J. (1981). *Biochemistry*, **20**, 2361.
41. Fredrikson, G., Nilsonn, S., Olsson, H., Björck, L., Åkerström, B., and Belfrage, P. (1987). *J. Immunol. Methods*, **97**, 65.
42. Gomi, H., Hozumi, T., Hattori, S., Tagawa, C., Kishimoto, F., and Björck, L. (1990). *J. Immunol.*, **144**, 4046.
43. Füglistaller, P. (1989). *J. Immunol,. Methods*, **124**, 171.
44. Pharmacia (1994). *Monoclonal antibody purification handbook*.
45. Juarez-Salinas, H. and Ott, G. S. (1987). United States Patent no. 4,704,366.
46. Beyzavi, K. and Clift, R. (1994). *Genet. Engng News*, **March**, 17.
47. Forsgren, A., Ghetie, V., Lindmark, R., and Sjöquist, J. (1982). In *Staphylococci and staphylococcal infections* (ed. C. S. F. Eastman and C. Adlam), Vol. 2, pp 429–80. Academic Press, London.
48. Francis, R., Bonnerjea, J., and Hill, C. R. (1990). In *Separations for biotechnology* (ed. D. L. Pyle), Vol. 2, pp 491–8. Elsevier, London.
49. Ohlson, S. and Niss U. (1988). *J. Immunol. Methods*, **106**, 225.
50. Hodgson, J. (1990). *Bio/Technology*, **8**, 864.
51. Savin, J. (1990). *The value of antibody engineering technology to the UK*. Centre for Exploitation of Science and Technology, London, UK.
52. Khatter, N., Matson, R. S., and Ngo, T. (1991). *Amer. Lab.*, **23**, 32.
53. Clonis, Y. D. (1987). In *Reactive dyes in protein and enzyme technology* (ed. Y. D. Clonis, T. Atkinson, C. J. Bruton, and C. R. Lowe), pp. 33–50. Macmillan Press, London.
54. Squire, R. S., Greenwood, J. V., and Stanford, L. M. (1993). *J. Reprod. Immunol.*, **23**, 207.
55. Godfrey, M. A. J., Kwasowski, P., Clift, R., and Marks, V. (1994). In *Separations for biotechnology* (ed. D. L. Pyle), Vol. 3, pp. 200–206. Royal Society of Chemistry, Cambridge, UK.

56. Porath, J., Maisano, F., and Belew, M. (1985). *FEBS Lett.*, **185**, 306.
57. Lihme, A. L. and Heegaard, M. H. (1991). *Anal. Biochem.*, **192**, 64.
58. Tiselius, A., Hjertén, S., and Levin, O. (1956). *Arch. Biochem. Biophys.*, **65**, 132.
59. Cummings, L. J., Ogawa, T., and Tunón, P. (1994). In *Separations for biotechnology* (ed. D. L. Pyle), Vol. 3, pp. 134–40. Royal Society of Chemistry, Cambridge, UK.
60. Cummings, L. J. (1994). *Genet. Engng News*, **14**, 7.
61. Lu, J., Thiel, S., Wiedemann, H., Timpl, R., and Reid, K. B. M. (1990). *J. Immunol.*, **144**, 2287.
62. Hortin, G. L. and Trimpe, B. L. (1990). *Anal. Biochem.*, **188**, 271.
63. Axen, R., Porath, J., and Ernback, S. (1967). *Nature*, **214**, 1302.
64. Lowe, C. R. (1979). In *Laboratory techniques in biochemistry and molecular biology* (ed. T. S. Work and E. Work), Vol 7, pp. 269–518. Elsevier, London.
65. Inman, J. K. and Dintzis, H. M. (1969). *Biochemistry*, **8**, 4070.
66. Hortin, G. L. and Trimpe, B. L. (1990). *Anal. Biochem.*, **188**, 271.
67. Kabir, S. (1993). *Comp. Immunol. Microbiol. Infect. Dis.*, **16**, 153.
68. Loomes, L. M., Stewart, W. W., Mazengerara, R. L., Senior, B. W., and Kerr, M. A. (1991). J. Immunol. Methods, **141**, 209.
69. Henniker, A. J. and Bradstock, K. F. (1993). *Biomed. Chromatogr.*, **7**, 121.
70. Eliasson, M., Olsson, A., Palmcrantz, E., Wiberg, K., Inganäs, M., Guss, B., Lindberg, M., and Uhlén, M. (1988). *J. Biol. Chem.*, **263**, 4323.
71. Åkerström, B., Nielsen, E., and Björck, L. (1987). *J. Biol. Chem.*, **263**, 13388.
72. Morgan, M. R. A., Brown, P. J., Leyland, M. J., and Dean, P. D. G. (1978). *FEBS Lett.*, **87**, 239.
73. Ngo, T. T. and Khatter, N. (1992). *J. Chromatogr.*, **597**, 101.

7

Designer affinity purifications of recombinant proteins

SATISH K. SHARMA

1. Introduction

During the last five years, tailoring recombinant proteins to facilitate their purification has proven to be a very useful strategy in recombinant DNA technology. DNA encoding peptide or protein tags (also called, among other names, affinity tails, cleavable linkers, and marker sequences) is attached to the gene of interest at its 5′ or 3′ end. The resulting gene fusions are expressed in a host and the encoded recombinant protein can be isolated from contaminating host proteins based on properties of the engineered tag. A schematic representation of this genetic approach, called designer affinity purification, is shown in *Figure 1*.

Conceivably, the same purification methodology could be utilized for many different recombinant proteins and their analogues. This would obviate the need to devise and fine-tune purification protocols for each individual protein. It should be noted that the purified protein thus obtained still contains the 'affinity tag' which may be undesirable in some cases.

Basically, affinity tags can be classified into two categories depending upon their size and end use:

(a) temporary tags, which are generally large and thus could adversely affect the desired protein; they include:
   - β-galactosidase (116 kDa)
   - glutathione transferase (26 kDa)
   - Protein A (30 kDa)
   - maltose-binding protein (26 kDa)
   - cellulose-binding domain (100 amino acids)
   - Pinpoint™ (Promega) (13 kDa biotinylated fusion protein)

(b) permanent tags, which are small (<1 kDa) and unlikely to have a detectable effect on the structure and function of the protein; they include:
   - an oligopeptide (0.70 kDa)

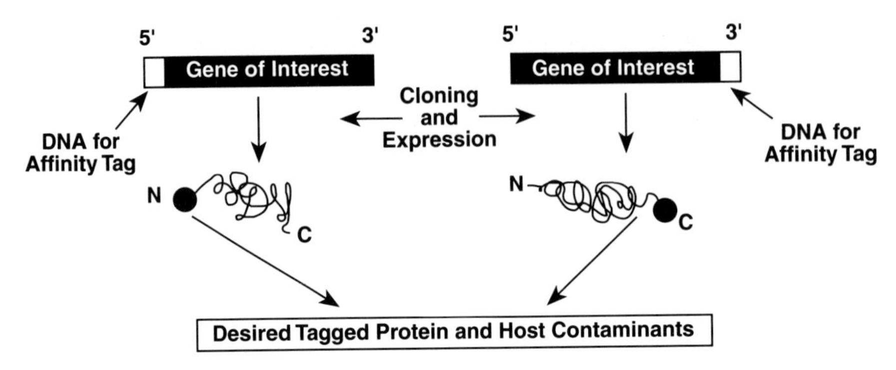

Figure 1. An illustration of genetically engineered tags for recombinant protein purification. •, affinity tag attached to the desired protein at its N- or C-terminus.

- hexahistidine (0.70 kDa)
- an 8-residue peptide containing alternating histidines (0.90 kDa)
- Flag™ peptide (Eastman Kodak) (8 amino acids)
- Strep Tag (10 amino acids)

'Temporary' tags often have to be removed from the protein of interest, either to restore function, increase solubility, avoid interference during assays or to prevent antigenicity if the protein is destined for therapeutic use. Tag removal can often be a very challenging process and, where absolutely required, may be the technical bottleneck despite the usefulness of the tag for purification of the desired protein. To address this impediment, a number of chemical and enzymatic techniques have been reported for the site-specific hydrolysis of recombinant proteins engineered with an affinity tag (see *Table 1*). Specific cleavage of recombinant proteins with enzymes is preferable to chemical methods. The enzymatic approach has several advantages:

- enzymatic cleavage sites occur less frequently in proteins and are more selective;
- enzymatic cleavage conditions are mild and thus compatible with more target proteins;
- enzymatic cleavage lacks undesirable side reactions.

However, enzymatic methods are not always practicable and sometimes their selectivity is dependent upon the sequences on each side of the cleavage site. This chapter is not intended to address practical aspects of tag removal following a designer affinity purification. Interested readers can refer to a comprehensive and illuminating review on this subject (1).

Designer affinity purifications have been used to obtain recombinant proteins based on the bioaffinity (2), immunoaffinity (3), ion-exchange (4), hydrophobic (5), covalent (6), and metal-chelating (7–9) properties of the

Table 1. Commonly used methods for chemical and enzymatic cleavage of fusion proteins[a]

Cleavage	Specificity
Chemical	
Cyanogen bromide	-Met- ↓ -X-
Enzymatic (1)	
Factor Xa	Ile-Glu-Gly-Arg- ↓ -X
Enterokinase	$(Asp)_4$-Lys- ↓ -X
Thrombin	Gly-Val-Arg-Gly-Pro-Arg- ↓ -X
V8 protease	Gly-Ser-Val-Asp-Glu- ↓ -X

[a]In some cases the specificity is dependent upon amino acids on the right-hand side of the cleavage site (arrow). For more details and specific references, see ref. 1.

engineered tags. Some companies (e.g. Qiagen) now market purification technologies based on engineered affinity tags.

In the approaches listed above, bioaffinity and metal affinity are most widely used for the selective adsorption of engineered recombinant proteins. Specifically, the most common bioaffinity system utilizes glutathione S-transferase (GST) fusion for purification on glutathione–Sepharose 4B (Pharmacia Biotech). On the other hand, a short peptide fusion with affinity for an immobilized metal ion is the basis for metal affinity purifications. The versatility of GST and metal-chelating fusions is vouched for by their routine use, as documented by publications from many molecular biology and biochemistry laboratories. Therefore, this chapter focuses on recombinant proteins containing metal chelate and GST fusions.

2. Immobilized metal affinity chromatography (IMAC)

2.1 Principles

The IMAC concept was originally introduced by Porath and co-workers (30,31). In solution, protein precipitation can occur because a single protein molecule can bind several metal ions which in turn can cross-link with other protein molecules. These molecular events can be decreased simply by coupling the metal ion to a solid support (30,31). Thus, in IMAC the metal ion is restrained in a coordination complex where it still retains significant affinity towards macromolecules. The idea is to optimize protein–metal ion interactions in order to prevent protein aggregation, thereby allowing high protein recovery while achieving significant purification. The basic principles underlying this technique have been described elsewhere (32,33). A schematic of IMAC during purification of a recombinant protein from a crude cell extract is shown in *Figure 2*.

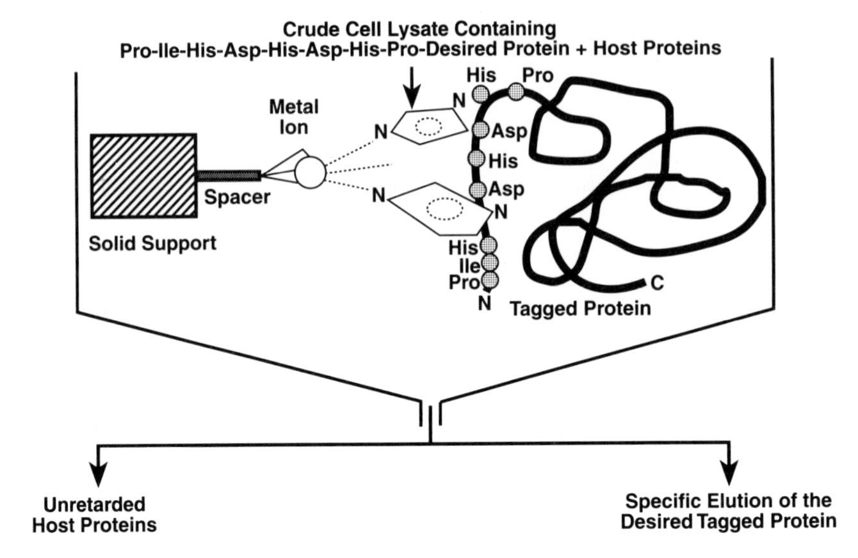

Figure 2. A schematic view of IMAC showing coordination of the imidazole groups of the desired tagged protein with the metal ion immobilized to a solid support.

IMAC systems have three basic components *(Figure 2)*: an electron donor group (e.g. histidine), a solid support, and a metal ion. The most commonly used solid supports for IMAC purification are iminodiacetic acid–Sepharose (IDA–Sepharose) (also known commercially as Fast Flow Chelating Sepharose (Pharmacia)) and nitrilotriacetic acid–Sepharose (NTA–Sepharose). In theory, a variety of metal ions with a high affinity for histidine ligands can be utilized in IMAC. Of these, Ni^{2+} is the most widely used metal ion in the purification of recombinant proteins engineered with histidine-containing IMAC tags. This is due to the fact that such IMAC tags seem to have very high affinity for the immobilized Ni^{2+}. *Table 2* summarizes the distinct

Table 2. Comparison of IMAC with immunoaffinity chromatography for the purification of tagged recombinant proteins

Characteristic	IMAC	Immunoaffinity
Specificity	high	high
Ligand availability	easy to obtain	relatively difficult
Ligand stability	high	low
Protein binding	high	low
Concentration of purified protein	high	low
Binding conditions	denaturants acceptable	denaturants unacceptable
Elution conditions	mild	often harsh
Scale up	easy	relatively difficult
Expense	low	high

Table 3. Recent examples of recombinant proteins purified by immobilized metal affinity chromatography (IMAC)

Protein	Engineered histidines	Reference
HIV-1 RT	His-Asp-His-Asp-His	10
β-Galactosidase	His-Asp-His-Asp-His	10
p15 HIV RNase H	His-Asp-His-Asp-His	11
p51 HIV-1 RT	His-Asp-His-Asp-His	12
HIV-1 RT	C-terminal $(His\text{-}Glu)_4$	13
HIV-1 protease	$(His)_3$	14
Antibody fragments	$(His)_5$	15
Citrate synthetase	$(His)_5$	16
Disulfide isomerase	$(His)_6$	16
E. coli RNase H	$(His)_6$	17
Thymidine kinase	$(His)_6$	18
p24 HIV-1 protein	$(His)_6$	19
Transcription factor	$(His)_6$	20
Mouse FKBP-52	$(His)_6$	21
Human FKBP-52	$(His)_6$	22
Serum response factor	$(His)_6$	23
HIV-1 RT	$(His)_6$	9, 24
Galactose dehydrogenase	$(His)_6$	25
Yeast pre-pro-α-factor	$(His)_6$	26
HIV-1 Tat	$(His)_6$	27
Growth hormone releasing factor	$(His)_6$	28
Chick ferredoxin	$(His)_6$	29

characteristics of IMAC. In most of these respects IMAC is superior to immunoaffinity chromatographic methods for recombinant proteins engineered for purification. Although no tag is universally applicable, the general use of histidine-containing, metal-chelating-tags, referred to here as IMAC tags, is gaining significant popularity in this field. This is supported by a number of recent examples summarized in *Table 3*. In addition to the advantages listed in *Table 2*, the IMAC tags are considered permanent tags as they seem not to affect the properties of the desired protein.

2.2 General considerations in designing IMAC-based purification tags

The basic idea of using engineered IMAC tags is to provide the highest selectivity in the protein of interest for an immobilized metal ion. Thus, when designing an IMAC tag, the following points must be kept in mind before engineering a tag within the protein of interest:

- accessibility and affinity of the IMAC tag for the chosen immobilized metal ion
- amino acid sequences surrounding the IMAC tag
- hydrophilicity of the IMAC tag

Table 4. Imidazole elution of recombinant proteins engineered for purification by IMAC[a]

IMAC-specific tag	Protein	Elution (mM imidazole)	Reference
PIHHHHHHPFHLVIH	HIV-1 RT	300	9
MPIHDHDHPFHLVIHS	HIV-1 RT	100	10
PIHDHDHPFHGYQ	p15 RNase H	40	11
PIHDHDHPFHLVIHS	p51 RT	100	12
MEHEHEHEHE	HIV-1 RT	60–80	13
NGHDHDHDHDHL (C-terminal)	HIV-1 RT	50–70	13
MTMITPSSHHHHHH	Upstream stimulatory factor	2000	20
MSHHHHHHGEFPG	Serum response factor	80	23

[a]Reproduced from reference 13 with permission of the publisher. Unless stated otherwise, all the tags are at the N-terminus.

The degree of selectivity and affinity for an immobilized metal ion can be influenced by sequences surrounding the IMAC tag *(Table 4)*. Moreover, selection of the engineering site on the protein of interest is critical to ensure that the activity of the protein is not diminished by the presence of the tag. For example, if it is known that modification of the N-terminal site is not tolerated by the protein of interest, the tag should be placed at its C-terminus instead.

2.3 Selection of the metal ion

In proteins, the amino acid histidine is the prime candidate for coordination with immobilized transition metal ions such as Cu^{2+}, Co^{2+}, Ni^{2+}, and Zn^{2+} (30–33). Several other factors can influence the interaction between the metal ion and the histidine ligand:

- the effect of adjacent charged residues on the pK_a of histidines
- the steric characteristics of the bulky side-chains
- the overall charge of the protein of interest
- the pH of the protein-containing solution

Thus, it is important to determine the best metal ion for binding and elution of proteins containing IMAC tags. Two important considerations are that the host proteins must have low affinity for the metal ion, and the protein of interest must have high affinity due to the engineered tag.

Figure 3 shows an example of IMAC for the purification of the enzyme human immunodeficiency virus type 1 (HIV-1) reverse transcriptase (RT) engineered with a histidine-containing tag; Pro-Ile-His-Asp-His-Asp-His-Pro-Phe-His-Leu-Val-Ile-His-Ser (34). The effect of Zn^{2+}, Cu^{2+}, and Ni^{2+} on binding and elution of this enzyme was examined. The engineered HIV-1 RT

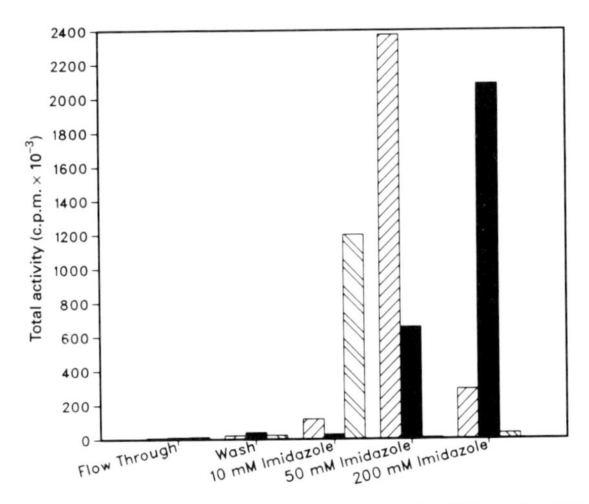

Figure 3. Effect of various immobilized metal ions [Zn^{2+} (▨), Ni^{2+} (■), Cu^{2+} (◪)] on IMAC of HIV-1 RT engineered with an N-terminal affinity tag (Pro-Ile-His-Asp-His-Asp-His-Pro-Phe-His-Leu-Val-Ile-His-Ser) containing alternating histidines (10). Elution was carried out with 10 mM, 50 mM, and 200 mM imidazole. Flow-through, wash, and eluted pools were tested for HIV-1 RT activity indicated as total activity (c.p.m.). Reproduced from ref. 10 with permission.

did not bind to an immobilized Cu^{2+} column, while binding to Zn^{2+} was considered moderate. However, a strong retention was observed when a solution of the HIV-1 RT was passed through a Ni^{2+} column. From these screening experiments, it was concluded that Ni^{2+} is the best metal ion for HIV-1 RT endowed with this tag. Although metal-ion leaching during IMAC at neutral pH is very low, the effect of a metal ion on the enzymatic activity of the desired protein should also be taken into consideration. It should be noted that an apoenzyme may scavenge the metal ion during IMAC and emerge as a metalloprotein (33).

2.4 Choice of the IMAC tag

The majority of IMAC applications have involved the hexahistidine tag, as shown in *Table 3*. If the purpose of the IMAC tag is only for affinity purification, a sequence of six neighbouring histidines is a very good choice. However, a hexahistidine sequence may not be suitable for applications such as tag-based immunodetection of recombinant proteins. Thus, in some situations it might be important to explore other histidine-containing tags which could be utilized for both purification and immunodetection. A metal-binding peptide containing alternating histidines for purification and immunodetection has been reported (35). In such cases, it is important that the alternating amino acids be hydrophilic residues known to be exposed and to contribute to antigenic epitopes. In turn, such sequences would be exposed for recognition by the antibody. Thus, a short stretch of alternating histidines is recommended if, in addition to purification, one also wants to exploit the tag for immunodetection applications (see Section 4).

2.5 Selection of binding and elution conditions

The selection of an appropriate buffer, and binding and elution conditions, is crucial for the success of IMAC. Although Tris buffer has been found to be very useful in most of the reported applications, other buffers (including Mops, Mes, and Hepes) have been recommended. The pH optimum depends upon the choice of the metal ion and, therefore, buffer. Generally, a pH of 8 works well for immobilized Ni^{2+}. Under certain conditions, electrostatic interactions can occur simultaneously with coordinate bonds and, therefore, the use of 0.5–1 M NaCl is recommended to prevent ion-exchange effects during IMAC. IMAC is compatible with non-ionic detergents, 3 M urea, and 6 M guanidinium HCl (8,36). The temperature should be chosen based on the stability of the protein of interest.

There are two well established modes of eluting bound proteins from an IMAC column:

- elution by lowering the pH
- elution by using competitive counter-ligands (e.g. imidazole)

All the examples given in this chapter (see Section 3) use alternating histidine tags and the mode of elution is by competition with imidazole. Generally, the choice of the elution method and its mode (stepwise or gradient) should be determined experimentally on a case-by-case basis. A list of imidazole elution conditions used to obtain purified proteins by IMAC is also shown in *Table 4*. This table should serve as a guide for the approximate imidazole concentration needed to achieve purification of a recombinant protein engineered with a given IMAC sequence (see *Table 3*).

3. Methodology for cloning, expression, and purification of proteins engineered with an alternating histidine tag

3.1 Engineering the IMAC tag

In this section, the methodology for creating constructs suitable for designer affinity purifications is briefly described for each specific example. Genes coding for β-galactosidase, the p51 polymerase domain of HIV-1 RT, and HIV-1 RT were constructed and cloned into *Escherichia coli* expression vectors as described (12,13,34). The recombinant β-galactosidase and p51 polymerase domain proteins were designed to possess the tag Pro-Ile-His-Asp-His-Asp-His-Pro-Phe-His-Leu at the N-terminus. The p51 polymerase domain also contained the additional sequence Val-Ile-His-Ser as part of the renin cleavage site (12).

The expression plasmids for β-galactosidase and the p51 polymerase domain of HIV-1 RT, engineered with an N-terminal IMAC tag, are shown in *Figure 4a* and *b*, respectively. Other details regarding the molecular biology

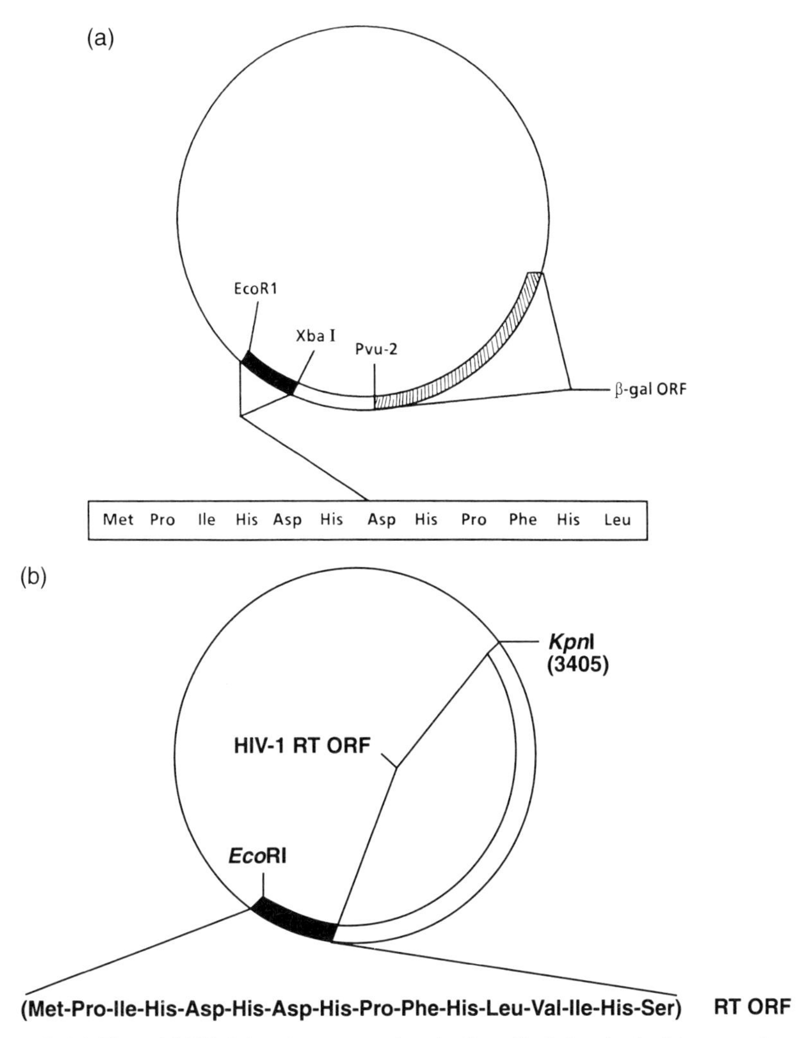

Figure 4. (a) Plasmid DE 6 for the expression in *E. coli* of β-galactosidase engineered with alternating histidines (34). Key: solid bar, synthetic oligonucleotide; white bar, 33 nucleotides from pUC18; hatched bar, β-galactosidase ORF; solid line, remainder of plasmid. Reproduced from ref. 34 with permission. (b) Plasmid DE-11 used for expression of p51 polymerase domain (amino acid residues 1–427) of HIV-1 RT. Key: solid bar, synthetic oligonucleotide encoding the N-terminal linker sequence containing the IMAC affinity tag; white bar, HIV-1 RT ORF (residues 1–427); solid line, pKK223-3 vector with ptac promoter. Reproduced from ref. 12 with permission.

of these constructs are available (12,13,34). The construction of the plasmid expression vector encoding HIV-1 RT with a C-terminal IMAC tag is reported elsewhere (13). This example was selected to show how HIV-1 RT can be purified based on an IMAC tag present at the C-terminus (HIV-1 RT-Asn-Gly-(His-Asp)$_4$-His-Leu).

Prior to transformation, the cloned DNA sequence of each plasmid was confirmed by dideoxy DNA sequencing with [^{32}P]dATP (13). The plasmids were then transformed into the JM109 strain of *E. coli*, as detailed in *Protocol 1*.

Protocol 1. Transformation and expression of tagged recombinant proteins in *E. coli*

Equipment and reagents

- Spectrophotometer (e.g. Bausch and Lomb, Spectronic 2000)
- Centrifuge (RC 5C and SS-34 rotor, Dupont) or their equivalents
- Microprocessor shaker bath (e.g. Lab-Line)
- CO_2 incubator-shaker (e.g. Lab-Line)
- NS-85 medium (9) supplemented with 1% yeast extract, 50 μg/ml thiamin, and 100 μg/ml ampicillin in 400 ml batches in 2 litre flasks
- Agar plates
- Ampicillin (Sigma)
- Isopropyl β-D-thiogalactopyranoside (Gibco-BRL)
- Competent JM109 *lacI^Q E. coli* cells prepared by the $CaCl_2$ method (ref. 37)
- Recombinant plasmid DNA containing insert encoding the desired protein
- L broth (Gibco-BRL)
- LB/Amp (L broth containing 100 μg/ml ampicillin)

Methods

1. Transform recombinant plasmid DNAs into competent JM109 *lacI^Q E. coli* cells using standard procedures (13,34,37).
2. Perform clonal selection by growing clones at 37 °C on LB/Amp plates.
3. Grow 2 ml of liquid 'seed' cultures of the selected clones in LB/Amp. Add 80% glycerol at a ratio of 2.3:1 (v/v) and store at −80 °C.
4. For every 400 ml batch of NS-85 medium inoculate one 5 ml primary seed culture in LB/Amp medium of selected clones. Grow cells at 37 °C, overnight.
5. Inoculate 400 ml batches of NS-85 medium with primary seed cultures. Grow these cultures at 37 °C with continuous shaking.
6. Induce the cells by adding isopropylthio-β-D-galactoside to 1 mM final concentration when the absorbance at 600 nm is approximately 0.4 units.
7. Grow cells for another 4 h at 37 °C.
8. Harvest the cells by centrifugation at 8000*g* for 10 min at 4 °C. Discard the supernatant.
9. Weigh the cell paste and further process it by purification of the desired protein by IMAC (see Section 3.2).

3.2 Purification of recombinant proteins containing an IMAC tag

The commercially available Chelating Sepharose (Fast Flow) contains immobilized iminodiacetic acid (IDA). This resin is used for immobilizing metal ions (e.g. Ni^{2+}), and then for purifying the tagged protein (see *Figure 2*). The expression of tagged proteins in *E. coli* is described in *Protocol 1*. *Protocols 2–4* deal with the purification methodology. Specific examples of IMAC purification of histidine-containing tagged proteins are described in Section 3.3.

Protocol 2. Preparation of immobilized metal affinity columns for protein purification

Equipment and reagents

- IDA–Sepharose (Pharmacia, code 17-0575-02)
- Milli-Q (Millipore) or double-distilled water
- Glass columns (1.5 cm × 8 cm and 1.7 cm × 15 cm)
- Peristaltic pump (e.g. P-1, Pharmacia)
- 50 mM EDTA (disodium) adjusted to pH 8 with 10 M NaOH
- Equilibration buffer: 20 mM Tris–HCl (electrophoresis grade) 0.5 M NaCl, 1 mM PMSF, 1 mM benzamidine hydrochloride, 10 mg/litre leupeptin, 10 mg/litre beastatin hydrochloride, 10 mg/litre aprotinin pH 8.
- 50 mM nickel sulfate in Milli-Q water
- 0.2 M NaOH

Methods

1. Using a sintered glass filter, wash the IDA–Sepharose with Milli-Q (Millipore) water. **Warning:** Do not let the resin dry on the filter.
2. Form a slurry by adding water and pour it into a glass column to a volume of 6 ml gel (1.5 cm × 8 cm) or 15 ml gel (1.7 cm × 15 cm).
3. Allow the gel to settle by closing the column valve at the bottom.
4. Open valve and using a peristaltic pump at a flow rate of 1 ml/min, wash the column with 5 column volumes each of 50 mM EDTA pH 8.0, 0.2 M NaOH, and Milli-Q water.
5. Charge the column with five column volumes of 50 mM nickel sulfate.
6. Equilibrate Ni^{2+} charged column (step 5) with five column volumes of the equilibration buffer.

Protocol 3. Preparation of cell free crude *E. coli* extracts for IMAC

Equipment and reagents

- French pressure cell press (e.g. American Instrument Co.)
- French pressure cell (e.g. Aminco)
- Centrifuge (RC 5C & SS-34 rotors (Dupont) or equivalent)
- 2 M Tris–HCl, pH 8.0
- IMAC column (see *Protocol 2*). Use 1.5 cm × 8 cm column for 2–5 g of cell paste and 1.7 cm × 15 cm column for 5–10 g of cell paste
- Cell paste (from *Protocol 1*, step 9)
- Equilibration buffer (see *Protocol 2*)

Protocol 3. *Continued*

Method

1. Suspend 4–10 g of the cell paste in 5–10 vols (5–10 ml/g) of equilibration buffer.
2. Stir the suspension at 4 °C for at least 1 h, preferably overnight.
3. Cool all components of the French pressure cell at 4 °C overnight.
4. Break the cells at 1100 bar by passing the suspension through the pre-cooled French press.
5. Repeat step 4 twice more to ensure breakage of all the cells. **Warning:** Between successive applications of pressure, cool the suspension and the French pressure cell for 10 min at 4 °C.
6. Stir the total cell lysate for 1 h at 4 °C.
7. Centrifuge for 1 h (10 000*g*) at 4 °C.
8. Discard the pellet and determine the volume and the pH of the cell-free extract.
9. Adjust the pH to 8.0 with 2 M Tris–HCl, pH 8.0.
10. Apply the supernatant on to the Ni^{2+} charged IMAC column at a flow rate of 0.5 ml/min using a peristaltic pump.

Protocol 4. IMAC at 4 °C: binding and elution of tagged proteins

Equipment and reagents

- Spectrophotometer (e.g. Bausch and Lomb, Spectronic 2000)
- Fraction collector (e.g. Gilson)
- Peristaltic pump (P-1, Pharmacia)
- 20 mM Tris–HCl, 1 M NaCl pH 7.2
- IMAC column (prepared as in *Protocol 3*)
- 50 mM EDTA
- 0.2 M NaOH
- 50 mM nickel sulfate
- Imidazole (35, 50, 100, and 150 mM solutions in 20 mM Tris–HCl, 0.5 M NaCl pH 8.0)
- Equilibration buffer (see *Protocol 2*)

Method

1. Wash the column with equilibration buffer after application of the cell-free extract to the pre-equilibrated Ni^{2+}-bound IDA–Sepharose column (as in *Protocol 3*).
2. Read absorbance at 280 nm and continue washing until absorbance reaches zero to ensure complete removal of non-specifically bound contaminating proteins.
3. Elute unwanted, host-related bound proteins by washing the column with 20 mM Tris–HCl, 0.5 M NaCl, 35 mM imidazole, pH 8.0, buffered with HCl.

4. Elute the desired fusion protein using either:
   (a) a linear gradient of imidazole containing 15 ml each of 50 mM imidazole and 150 mM imidazole in 20 mM Tris–HCl, 0.5 M NaCl, pH 8.0 or
   (b) 30 ml of 100 mM imidazole in Tris buffer pH 8.0.
5. Remove the imidazole and traces of metal ion from the purified protein by overnight dialysis against 20 mM Tris–HCl, 1 M NaCl, pH 7.2 at 4 °C.
6. The purified protein samples thus obtained can be stored at −80 °C until further use. **Warning:** The purified proteins irreversibly precipitate out when stored at −20 °C or −80 °C without the dialysis step (step 5).
7. Regenerate the used IMAC column by successive washing it with five column volumes of:
   (a) 50 mM EDTA
   (b) 0.2 M NaOH
   (c) distilled water
   (d) 50 mM nickel sulfate

3.3 Specific examples of IMAC purification of recombinant proteins from crude *E. coli* extract

3.3.1 β-Galactosidase

A schematic representation of the recombinant β-galactosidase gene construct is shown in *Figure 4a*. The protein encoded by this construct contains the following N-terminal tag (note the alternating histidines): Pro-Ile-His-Asp-His-Asp-His-Pro-Phe-His-Leu-β-galactosidase. When a crude *E. coli* extract containing this recombinant protein was passed through an IMAC column, three protein peaks were eluted (*Figure 5a*). The first peak contained *E. coli* proteins with no affinity for the immobilized metal ion. The middle protein peak was eluted at 35 mM imidazole, and was devoid of the tagged β-galactosidase protein as determined by ELISA using antibodies against the IMAC tag (35). A sharp peak was obtained when 100 mM imidazole was used. This peak was confirmed to contain tagged β-galactosidase by ELISA using antibodies to the IMAC tag. The isolated tagged β-galactosidase protein was judged to be highly pure by SDS–PAGE, as shown in *Figure 5b*.

3.3.2 p51 polymerase domain of HIV-1 RT

This recombinant protein was expressed in *E. coli* and included a histidine-containing N-terminal tag: Pro-Ile-His-Asp-His-Asp-His-Pro-Phe-His-Leu-Val-Ile-His-Ser. *Figure 6* shows SDS–PAGE of the purified p51 polymerase

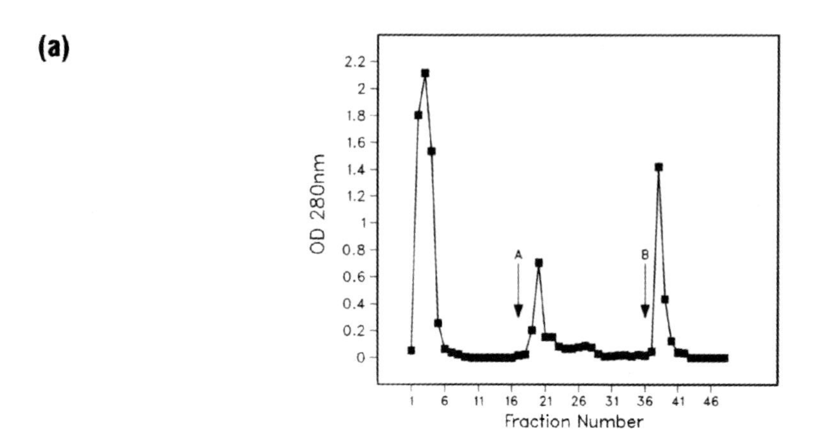

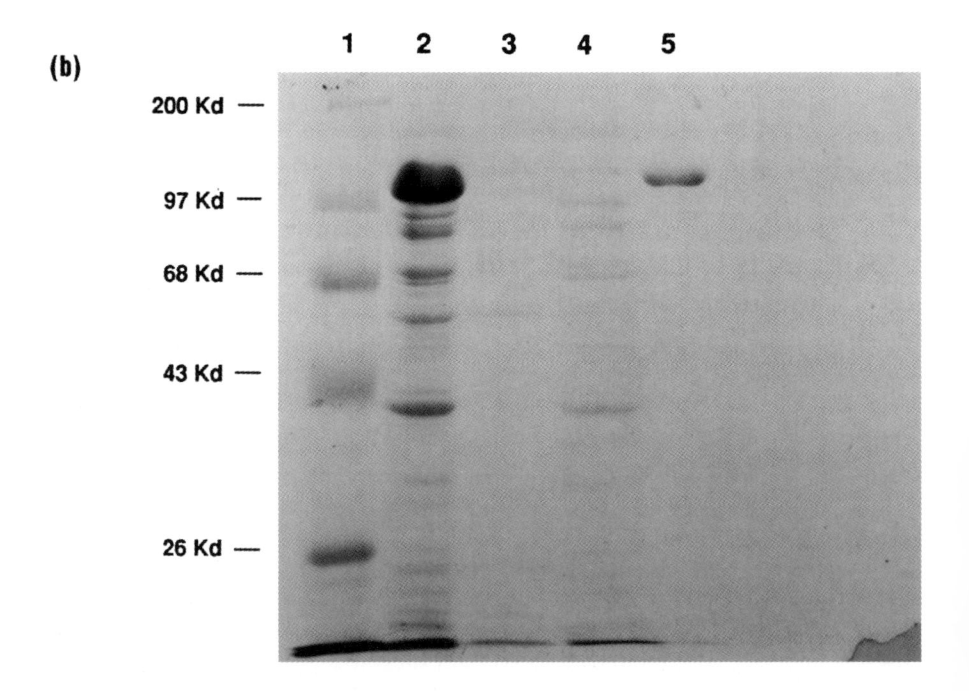

Figure 5. (a) IMAC of β-galactosidase containing an N-terminal IMAC tag. ■, OD at 280 nm. Arrow A, elution with 35 mM imidazole in buffer; arrow B, elution with 100 mM imidazole in buffer (see *Protocol 4*, step 4). (b) SDS–PAGE (12%) of protein-containing fractions from the elution profile shown in panel (a). Lane 1, molecular weight markers; lane 2, crude *E. coli* extract containing histidine-tagged β-galactosidase; lane 3, wash (pooled fractions 6–16); lane 4, 35 mM imidazole peak; lane 5, 100 mM imidazole peak. Reproduced from ref. 10 with permission.

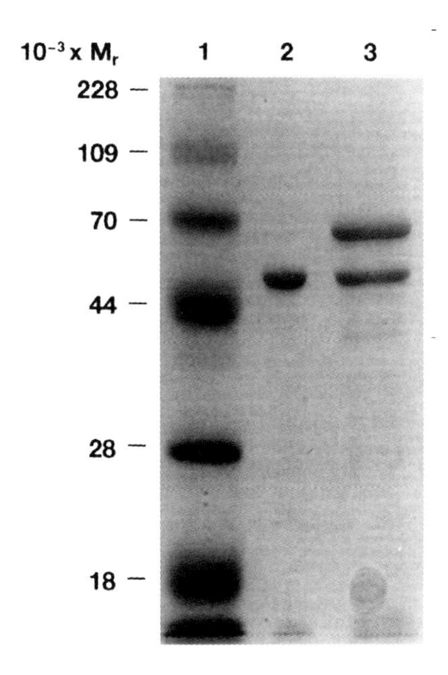

Figure 6. SDS–PAGE (12%) of IMAC-purified recombinant p51 HIV-1 RT. Lane 1, molecular weight markers; lane 2, purified p51 polymerase domain of HIV-1 RT; lane 3, p66/p51 heterodimeric HIV-1 RT which contains the hexahistidine N-terminal tag on both of its p51 kDa and p66 kDa subunits (24). Reproduced from ref. 12 with permission.

domain of HIV-1 RT obtained by single-step purification using this IMAC tag (12). It also shows, for comparison, a sample of purified heterodimeric HIV-1 RT (24) containing both the p66 and p51 polypeptides endowed with the commonly used hexahistidine IMAC tag (24). The results demonstrate that the IMAC tag associated with the p51 polymerase domain is as effective for IMAC purification as the widely used hexahistidine tag (see *Table 3*).

3.3.3 HIV-1 RT with a C-terminal IMAC tag

In this example, HIV-1 RT was engineered with a C-terminal tag (13) containing the sequence: Asn-Gly-(His-Asp)$_4$-His-Leu. An IMAC profile for this chimeric protein is shown in *Figure 7a*. As expected, the *E. coli* host proteins with no affinity for immobilized Ni^{2+} were not retained, while weakly retarded proteins were eluted in the presence of 35 mM imidazole. The desired protein was eluted using 100 mM imidazole. About 85% of the desired protein was recovered, as determined by activity analysis of this enzyme. *Figure 7b* shows the purity of the isolated protein analysed by SDS–PAGE. The presence of the C-terminal IMAC tag in the purified protein was confirmed by ELISA using antibodies (35) to the IMAC tag (see Section 4). This example shows that the C-terminal tag containing alternating histidines is accessible and can be used to purify recombinant proteins by

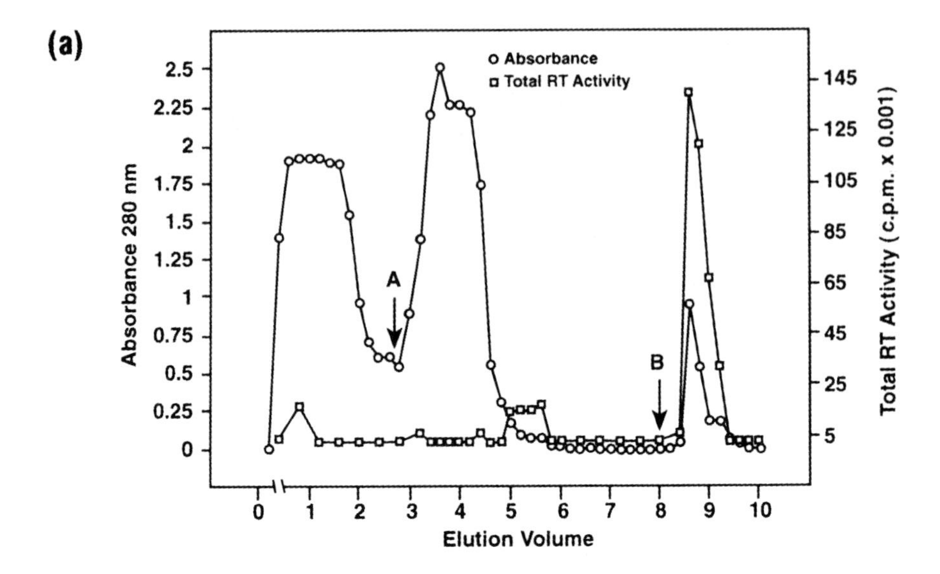

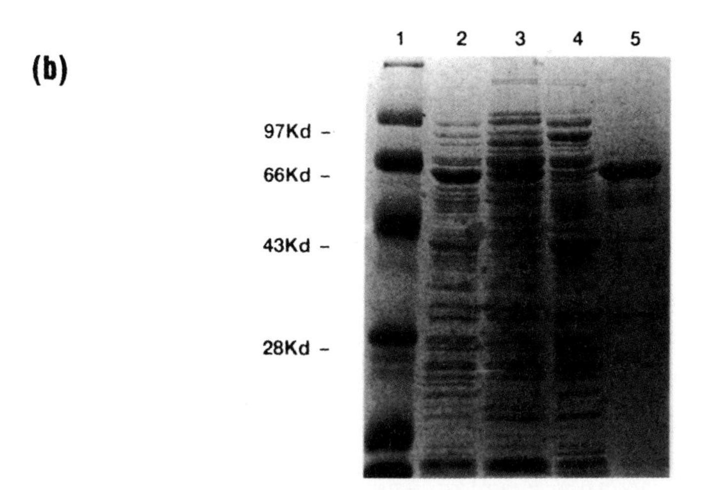

Figure 7. (a) Elution of HIV-1 RT containing the C-terminal IMAC tag from a Ni^{2+}-charged column. Crude *E. coli* extract (29 ml) was loaded on to a 1.5 cm × 8 cm column at a flow rate of 0.9 ml/min. One elution volume corresponds to 27.5 ml. At elution volume 2.8 (arrow A), 35 mM imidazole was included in the equilibration buffer. At elution volume 5, imidazole was omitted from the equilibration buffer. At elution volume 8 (arrow B), 100 mM imidazole was included in the equilibration buffer (see step 4 of *Protocol 4*). circles, absorbance at 280 nm; squares, RT activity. (b) A Coomassie stained SDS–polyacrylamide gel (11%) of protein-containing fractions from the IMAC column in panel (a). Lane 1, molecular weight markers; lane 2, cell-free *E. coli* extract; lane 3, buffer wash; lane 4, 35 mM imidazole elution; lane 5, 100 mM imidazole elution. Reproduced from ref. 13 with permission.

IMAC. This is especially important in cases where the exact native sequence is required at the N-terminus for structural or functional reasons. The recovered RT activity results also demonstrate that the IMAC tag at the C-terminus of HIV-RT did not interfere with the enzymatic activity of the protein of interest. This is consistent with the view that such small tags are permanent and cleavage might not be always essential to restore function.

4. Immunodetection of recombinant proteins engineered with an IMAC tag

One potential obstacle that sometimes hampers the application of recombinant DNA technology is the difficulty of detecting recombinant proteins during expression and purification. This is especially troublesome with structural proteins lacking functional assays. Engineered affinity tags have further facilitated the isolation of these proteins by acting as detection devices. One such specific example is the use of anti-peptide antibodies for immunodetection and immunoaffinity purification of recombinant proteins engineered with an immunogenic peptide (3). Thus, for an IMAC-specific tag to be useful in such applications, it must be immunogenic. Although hexahistidine fusions are extremely useful in IMAC purification (see *Table 3*), the tag itself is poorly immunogenic, so it seems difficult to develop antibodies against this tag for immunodetection.

A tagging approach that allows both IMAC purification and immunodetection of the tagged protein has been reported (35). This used the tag His-Asp-His-Asp-His-Pro-Phe-His-Leu, which is immunogenic in rabbits. The application of antibodies against this IMAC tag for immunodetection of tagged proteins also obviates the need for producing antibodies against each newly expressed protein. Tag-specific purification and immunodetection approaches should also be useful to distinguish between the recombinant protein and any homologous protein produced by host cells. In summary, the concept of designer IMAC purifications can also be applied to immunodetection of recombinant proteins based on the immunogenic properties of the IMAC-specific tag containing alternating histidines (35).

5. GST-based affinity purification of recombinant proteins

5.1 Introduction

A series of GST expression vectors, called pGEX, was first reported by Smith and Johnson (38). The desired protein can be expressed as a GST fusion protein which is purified from a crude extract by glutathione-affinity chromatography. As stated above, GST is a large affinity tag and, therefore, it is desirable to remove it from the purified fusion protein. This can be achieved

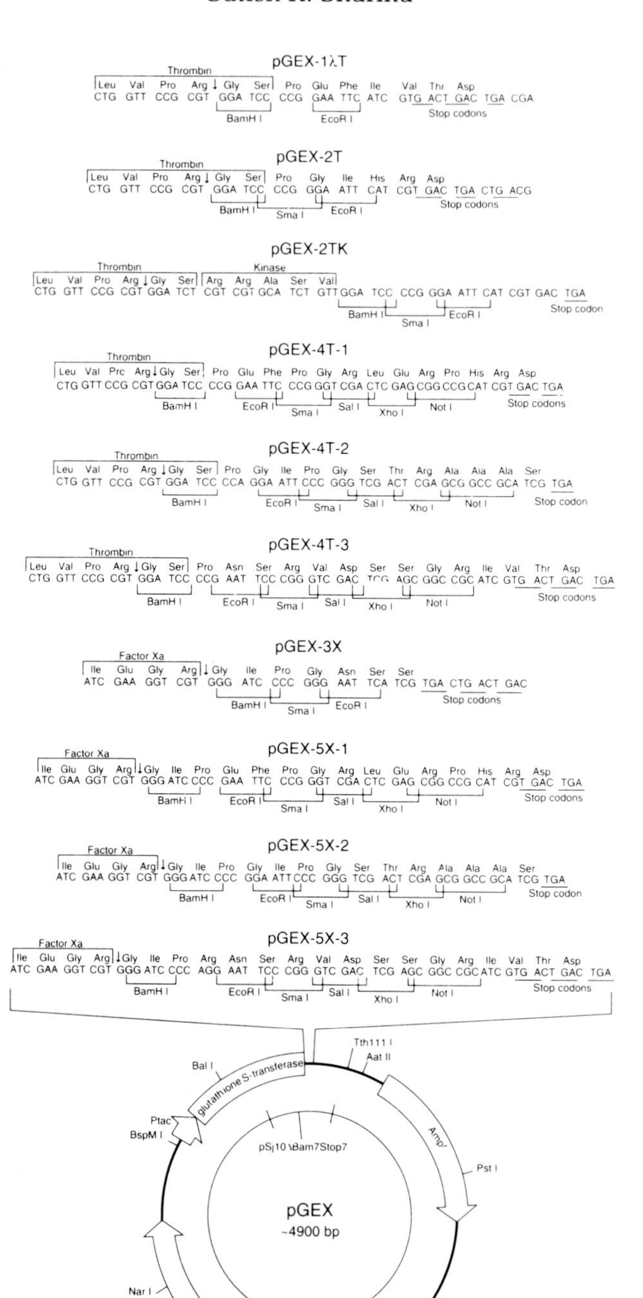

pGEX-1λT
Thrombin
Leu Val Pro Arg ↓ Gly Ser Pro Glu Phe Ile Val Thr Asp
CTG GTT CCG CGT GGA TCC CCG GAA TTC ATC GTG ACT GAC TGA CGA
BamH I
EcoR I
Stop codons
pGEX-2T
Thrombin
Leu Val Pro Arg ↓ Gly Ser Pro Gly Ile His Arg Asp
CTG GTT CCG CGT GGA TCC CCG GGA ATT CAT CGT GAC TGA CTG ACG
BamH I
Sma I
EcoR I
Stop codons
pGEX-2TK
Thrombin
Kinase
Leu Val Pro Arg ↓ Gly Ser Arg Arg Ala Ser Val
CTG GTT CCG CGT GGA TCT CGT CGT GCA TCT GTT GGA TCC CCG GGA ATT CAT CGT GAC TGA
BamH I
Sma I
EcoR I
Stop codon
pGEX-4T-1
Thrombin
Leu Val Pro Arg ↓ Gly Ser Pro Glu Phe Pro Gly Arg Leu Glu Arg Pro His Arg Asp
CTG GTT CCG CGT GGA TCC CCG GAA TTC CCG GGT CGA CTC GAG CGG CCG CAT CGT GAC TGA
BamH I
EcoR I
Sma I
Sal I
Xho I
Not I
Stop codons
pGEX-4T-2
Thrombin
Leu Val Pro Arg ↓ Gly Ser Pro Gly Ile Pro Gly Ser Thr Arg Ala Ala Ala Ser
CTG GTT CCG CGT GGA TCC CCA GGA ATT CCC GGG TCG ACT CGA GCG GCC GCA TCG TGA
BamH I
EcoR I
Sma I
Sal I
Xho I
Not I
Stop codon
pGEX-4T-3
Thrombin
Leu Val Pro Arg ↓ Gly Ser Pro Asn Ser Arg Val Asp Ser Ser Gly Arg Ile Val Thr Asp
CTG GTT CCG CGT GGA TCC CCG AAT TCC CGG GTC GAC TCG AGC GGC CGC ATC GTG ACT GAC TGA
BamH I
EcoR I
Sma I
Sal I
Xho I
Not I
Stop codons
pGEX-3X
Factor Xa
Ile Glu Gly Arg ↓ Gly Ile Pro Gly Asn Ser Ser
ATC GAA GGT CGT GGG ATC CCC GGG AAT TCA TCG TGA CTG ACT GAC
BamH I
Sma I
EcoR I
Stop codons
pGEX-5X-1
Factor Xa
Ile Glu Gly Arg ↓ Gly Ile Pro Glu Phe Pro Gly Arg Leu Glu Arg Pro His Arg Asp
ATC GAA GGT CGT GGG ATC CCC GAA TTC CCG GGT CGA CTC GAG CGG CCG CAT CGT GAC TGA
BamH I
EcoR I
Sma I
Sal I
Xho I
Not I
Stop codons
pGEX-5X-2
Factor Xa
Ile Glu Gly Arg ↓ Gly Ile Pro Gly Ile Pro Gly Ser Thr Arg Ala Ala Ala Ser
ATC GAA GGT CGT GGG ATC CCC GGA ATT CCC GGG TCG ACT CGA GCG GCC GCA TCG TGA
BamH I
EcoR I
Sma I
Sal I
Xho I
Not I
Stop codon
pGEX-5X-3
Factor Xa
Ile Glu Gly Arg ↓ Gly Ile Pro Arg Asn Ser Arg Val Asp Ser Ser Gly Arg Ile Val Thr Asp
ATC GAA GGT CGT GGG ATC CCC AGG AAT TCC CGG GTC GAC TCG AGC GGC CGC ATC GTG ACT GAC TGA
BamH I
EcoR I
Sma I
Sal I
Xho I
Not I
Stop codons
Tth111 I
Aat II
Bal I
glutathione S-transferase
Ptac
BspM I
pSj10 ∆Bam7Stop7
Amp^r
Pst I
pGEX
~4900 bp
Nar I
EcoR V
BssH II
lac I^q
p4.5
AlwN I
Apa I
BstE II
Mlu I
pBR322 ori

Figure 8. Some of the readily available GST fusion expression vectors. Details of the individual pGEX plasmids are also shown. This figure is reproduced with the kind permission of Pharmacia Biotech (Piscataway, New Jersey), and is taken from page 142 of their 1994 Molecular and Cell Biology catalogue.

Table 5. Potential problems and troubleshooting in the purification of GST fusion proteins

Problem	**Suggested solution**
No expression of the desired protein	Verify that the DNA insert and the pGEX vector are in the same reading frame (48).
Fusion protein is labile	Add protease-inhibitors to the cell-free extract as well as to the equilibration buffer as described in *Protocol 2*. Other possibilities include: use of protease-deficient host strains, growth of cells at lower temperature, or rapid processing of the crude extract at 4°C (48).
Insolubility of fusion protein	If the desired GST fusion produces an insoluble protein, it cannot be purified by direct binding to immobilized glutathione. This problem can sometimes be overcome either by growing cells between 10 and 25°C (49) or by solubilization of the cell lysate in detergents (50–53).

using vectors designed to contain specific proteolytic sites such as thrombin (pGEX-1, -2T) or factor Xa (pGEX-1, -3X). Since inception of the GST fusion approach (38), improvements have been made with regard to purification of insoluble proteins (39), rapid recovery of desired protein after fusion removal (40,41), vectors that enhance cleavage by thrombin (42,43), and new vectors that allow expression in yeast (44) or in insect cells (45–47).

5.2 Methodology for cloning, expression, and purification of proteins engineered with a GST tag

Figure 8 shows the GST-based, prokaryotic gene fusion expression vectors that are now commercially available. In addition to purification and fusion removal, some of these vectors allow detection of the fusion protein using either a biochemical or immunological assay. These expression vectors are compatible with most *E. coli* strains (48). The standard protocols for screening transformants and for protein expression (48) are essentially similar to the procedures described in *Protocol 1*. Purification of GST fusion proteins and troubleshooting (48) are described in *Protocols 5* and *Table 5* respectively.

Protocol 5. Purification of GST fusion proteins

Equipment and reagents

- Glutathione–Sepharose 4B (Pharmacia)
- Milli-Q water (Millipore) or equivalent
- Chromatographic column (1.5 cm × 15 cm)
- Fraction collector
- Peristaltic pump
- Spectrophotometer
- Apparatus for protein gel electrophoresis
- Reduced glutathione (5 mM in 50 mM Tris–HCl pH 8.0)
- Equilibration buffer (PBS): 150 mM NaCl, 16 mM Na_2HPO_4, 4 mM NaH_2PO_4 pH 7.3)

Method

1. Wash the glutathione–Sepharose 4B beads with 5–10 column volumes of PBS. **Warning:** Do not let the resin run dry at any time.
2. Pour the slurry into a suitably sized column and allow the gel to settle by closing the valve at the bottom.
3. Wash the column with equilibration buffer until the absorbance at 280 nm reaches zero and discard the eluent.
4. Apply the sample containing the desired fusion protein. Discard the eluent until the absorbance at 280 nm reaches zero.
5. Wash the column with five column volumes of equilibration buffer to ensure complete removal of non-specifically bound contaminating proteins.
6. Elute the desired fusion protein by competition with 5 mM reduced glutathione (GSH).
7. Analyse the purified fusion protein by 10% SDS–PAGE.
8. If removal of the GST fusion is desired, dialyse the purified protein against a buffer compatible with the enzymatic cleavage.

Note added in proof

While this work was in press an excellent review on affinity tags appeared (54).

Acknowledgements

I am thankful to Dr M. R. Deibel, K. B. Rank, and A. F. Vosters for helpful comments. Many thanks to all of my Pharmacia and Upjohn colleagues who have contributed to the development of IMAC-based affinity purifications and immunodetection procedures. *Figures 4b* and *6* are reproduced with permission from Portland Press, Ltd (United Kingdom); *Figures 3*, *4a*, *5* and *7* and *Table 4* are reproduced with permission from Academic Press (Orlando, FL). Many thanks to Pharmacia Biotech for permission to reproduce *Figure 8*. Thanks to all the publishers and the authors for their permission to use the published material. I am also grateful to Mrs Paula Lupina for her help during preparation of this chapter.

References

1. Carter, P. (1989). In *Protein purification: from molecular mechanisms to large-scale processes* (ed. M. R. Ladisch, R. C. Willson, C. C. Painton, and S. E. Builder), pp. 1181–93. American Chemical Society, Washington, DC.
2. Moks, T., Abrahmsen, L., Osterlof, B., Josephson, S., Ostling, M., Enfors, S. O., Persson, I., Nilsson, B., and Uhlen, M. (1987). *Bio/Technology*, **5**, 379.
3. Hopp, T. P., Pickett, K. S., Price, V. L., Libby, R. T., March, C. J., Ceretti, D. P., Urdal, D. L., and Conlon, P. J. (1988). *Bio/Technology*, **6**, 1204.
4. Sassenfeld, H. M. and Brewer, S. J. (1984). *Bio/Technology*, **2**, 76.
5. Persson, M., Bergstrand, M. G., Bulow, M., and Mosbach, K. (1988). *Anal. Biochem.*, **172**, 330.
6. Persson, M., Bulow, M. G., and Mosbach, K. (1990). *FEBS Lett.*, **270**, 41.
7. Smith, M. C., Furman, T. C., Ingolia, T. D., and Pidgeon, C. (1988). *J. Biol. Chem.*, **263**, 7211.
8. Hochuli, E., Bannwarth, W., Dobeli, H., Gentz, R., and Stuber, D. (1988). *Bio/Technology*, **6**, 1321.
9. Sharma, S. K., Evans, D. B., Vosters, A. F., McQuade, T. J., and Tarpley, W. G. (1991). *Biotech. Appl. Biochem.*, **14**, 69.
10. Vosters, A. F., Evans, D. B., Tarpley, W. G., and Sharma, S. K. (1992). *Protein Expr. Purif.*, **3**, 18.
11. Evans, D. B., Brawn, K., Deibel, M. R., Jr, Tarpley, W. G., and Sharma, S. K. (1991). *J. Biol. Chem.*, **266**, 20583.
12. Evans, D. B., Kezdy, F. J., Tarpley, G., and Sharma, S. K. (1993). *Biotech. Appl. Biochem.*, **17**, 91.
13. Sharma, S. K., Evans, D. B., Vosters, A. F., Chattopadhyay, D., Hoogerheide, J. G., and Campbell, C. M. (1992) In *Methods: A companion to Methods in enzymology* (ed. F. Arnold), Vol. 4, pp. 57–67. Academic Press, New York.
14. Leuthardt, A. and Roesel, J. L. (1993). *FEBS Lett.*, **326**, 275.
15. Skerra, A., Pfitzinger, I., and Pluckthun, A. (1991). *Bio/Technology*, **9**, 273.
16. Lindner, P., Guth, B., Wulfing, C., Krebber, C., Steipe, B., Muller, F., and Pluckthun, A. (1992) In *Methods: A companion to Methods in enzymology* (ed. F. Arnold), Vol. 4, pp. 41–56. Academic Press, New York.
17. Campbell, A. G. and Ray, D. S. (1993). *Proc. Natl Acad. Sci. USA*, **90**, 9350.
18. Franke, C. A. and Hruby, D. E. (1993). *Protein Expr. Purif.*, **4**, 101.
19. Cheynet, V., Bernard, V., and Mallet, F. (1993). *Protein Expr. Purif.*, **4**, 367.
20. Van Dyke, M. W., Sirito, M., and Sawadogo, M. (1992). *Gene,* **111**, 99.
21. Alnemri, E. S., Fernades-Alnemri, T., Nelki, D. S., Dudley, K., Dubois, G. C., and Litwack, G. (1993). *Proc. Natl Acad. Sci. USA*, **90**, 6839.
22. Peattie, D. A., Harding, M. W., Fleming, M. A., Decenzo, M. T., Lippke, J. A., Livingston, D. J., and Benasutti, M. (1992). *Proc. Natl Acad. Sci. USA,* **89**, 10974.
23. Janknecht, R., Martynoff, G. D., Jueren, L., Hipskind, R. A., Nordheim, A., and Stunnenberg, H. G. (1991). *Proc. Natl Acad. Sci. USA*, **88**, 8972.
24. Chattopadhyay, D. C., Evans, D. B., Deibel, M. R., Jr, Vosters, A. F., Eckenrode,

F. M., Einspahr, H. M., Hui, J. O., Tomasselli, A. G., Zurcher-Neely, H. A., Heinrikson, R. L., and Sharma, S. K. (1992). *J. Biol. Chem.*, **267**, 14227.
25. Lilius, G., Persson, M., Bulow, L., and Mosbach, K. (1991). *Eur. J. Biochem.*, **198**, 499.
26. Bush, G. L., Tassin, A.-M., Friden, H., and Meyer, D. I. (1991). *J. Biol. Chem.*, **266**, 13811.
27. Gentz, R., Chen, C. H., and Rosen, C. A. (1989). *Proc. Natl Acad. Sci. USA*, **86**, 821.
28. Basu, M., Dharm, E., Levine, J. F., Kramer, R. A., Crow, R. M., and Campbell, R. M. (1991). *Arch. Biochem. Biophys.*, **286**, 638.
29. Tang, C. and Henry, H. L. (1993). *J. Biol. Chem.*, **268**, 5069.
30. Porath, J., Carlsson, J., Olsson, I., and Belfrage, G. (1975). *Nature*, **258**, 598.
31. Porath, J. and Olin, B. (1983). *Biochemistry*, **22**, 1621.
32. Porath, J. (1988). *Trends Anal. Chem.*, **7**, 254.
33. Sulkowski, E. (1989). *BioEssays*, **10**, 170.
34. Evans, D. B., Tarpley, W. G., and Sharma, S. K. (1991). *Protein Expr. Purif.*, **2**, 205.
35. Evans, D. B., Vosters, A. F., Carter, J. B., and Sharma, S. K. (1992). *J. Immunol. Methods.*, **156**, 231.
36. Smith, J. S. and Roth, M. J. (1993). *J. Virol.*, **67**, 4037.
37. Sambrook, J., Fritsch, E. F., and Maniatis, T. (1989). *Molecular cloning: a laboratory manual*, 2nd edn, pp. 1.82–1.84. Cold Spring Harbor Laboratory Press, Cold Spring Harbor, NY.
38. Smith, D. B. and Johnson, K. S. (1988). *Gene*, **67**, 31.
39. Frangioni, J. V. and Neel, B. G. (1993). *Anal. Biochem.*, **210**, 179.
40. Abath, F. G. C. and Simpson, A. J. G. (1990). *Peptide Res.*, **3**, 167.
41. Grieco, F., Hay, J. M., and Hull, R. (1992). *BioTechniques*, **13**, 856.
42. Guan, K. and Dixon, J. E. (1991). *Anal. Biochem.*, **192**, 262.
43. Hakes, D. J. and Dixon, J. E. (1992). *Anal. Biochem.*, **202**, 293.
44. Mitchell, D. A., Marshall, T. K., and Deschenes, R. J. (1993). *Yeast*, **9**, 715.
45. Parker, L. L., Atherton-Fessler, S., and Piwnica-Worms, H. (1992). *Proc. Natl Acad. Sci. USA*, **89**, 2917.
46. Davies, A. H., Jowett, J. B. M., and Jones, I. M. (1993). *Bio/Technology*, **11**, 933.
47. Harris, M. and Coates, K. (1993). *J. Gen. Virol.*, **74**, 1581.
48. Smith, D. B. (1993). *Methods Mol. Cell Biol.*, **4**, 220.
49. Schein, C. H. and Noteborn, H. M. (1988). *Bio/Technology*, **6**, 291.
50. Blanc, S., Cerutti, M., Usmany, M., Vlak, J. M., and Hull, R. (1993). *Virology*, **192**, 643.
51. Frorath, B., Abney, C. C., Berthold, H., Scanarini, M., and Northemann, W. (1992). *BioTechniques*, **12**, 558.
52. Hartman, J., Daram, P., Frizzel, R. A., Rado, T., Benos, D. J., and Sorscher, E. J. (1992). *Biotechnol. Bioengng*, **39**, 828.
53. Bobek, L. A., Wang, X., and Levine, M. J. (1993). *Gene*, **123**, 203.
54. Nygren, Per-Åke., Ståhl, S., and Uhlen, M. (1994). *Trends Biotechnol.*, **19**, 184.

8

Affinity separations of cells

ERIK LUNDGREN

1. Introduction

There are more than 200 different major cell types in higher eukaryotes such as mammals. The isolation of each cell type with reasonable yield and high purity has become routine in many experimental systems as a part of a general reductionistic strategy. To analyse complex interactions in any system, we need to know the components involved and their properties.

Similar methods of cell separation are used in present and possible future clinical applications such as bone marrow transplantation, prenatal diagnosis, and gene therapy. This chapter will deal mainly with methods useful in practical experimental work, as the methods used in the clinic beyond those presented here are complicated by the need to apply Good Manufacturing Practice (GMP) and Good Laboratory Practice (GLP) regulations.

1.1 General strategy for separation of cells

Affinity methods are powerful tools for cell separation with high specificity, but are only part of a general strategy for cell separation. They are combined with other methods building on physical parameters such as sedimentation velocity, density, or adsorption to different surfaces (1). A successful separation strategy usually involves a combination of methods depending on what is being investigated and on the desired purity, cell number, and cell status after separation, or the analytical methods to be used in the ensuing studies. The availability of monoclonal antibodies and methods for physical isolation of antibody-coupled matrices have in a short time changed many established cell separation strategies in cell biology laboratories, as will be discussed in this chapter.

Affinity methods for cell separation rely on specific reagents, which are able to recognize structures expressed on cell surfaces. The reagents are fixed on solid supports. The physical forces needed to remove the cells of interest are stronger than those used to separate macromolecules. There are several reasons for this:

- Many cells occur in solid tissues, held together by adhesion molecules, tight junctions, etc. Therefore, they have to be detached before affinity separations.

Table 1. Examples of affinity separations of non-haematopoietic cells

Cell type	Affinity ligand[a]	Reference
Intestinal cell lines	biotinylated enterotoxin	13
Giant cells	mAb	32
Muscle cells	mAb/NCAM	17
Motor neurones	mAb/NGFrec	33
KB epithelial cell line	mAb/multidrug resistance	34
Endothelial cells	lectin	35
Platelets	collagen	36
Islet cells (pancreas)	mAb	37
Spermatozoa	mAb	38

[a]mAb, monoclonal antibody; NCAM, neuronal cell adhesion molecule; NGF, nerve growth factor.

- Even after suspension, cells adhere to each other or to contaminating macromolecules; this adhesion may be by non-specific adsorption (e.g. due to charge), or mediated by specific adhesion molecules.

Thus, the trend is to use methods like repeated rinsing, centrifugation, or strong magnetic fields rather than chromatographic methods. Another disadvantage of chromatographic methods for cell separation is that, due to the large size of cells (relative to macromoelcules), separation by chromatography is very slow and so there is a risk of microbial contamination.

Affinity methods are almost exclusively used for studies of cells in suspension; such cells may be derived from bone marrow or peripheral blood, for example, or they may be detached tissue culture cells. All detachment methods impose a limitation, as the methods used may disturb the molecules that are the targets for the affinity separations. Thus, proteolytic enzymes like trypsin or collagenase are known to strip surface molecules including important epitopes from cell surfaces. Mild methods for cell disaggregation based on EDTA are therefore often used (*Protocol 1*).

This chapter introduces the general principles for affinity separation of cells, mainly using as examples the separation of blood cells, which are an unusually complex mixture of many cell types. However, other types of mammalian cells can also be separated with these methods; some examples are summarized in *Table 1*. Affinity separation is not used for plant cells due to their thick cell walls. Protoplast separation is possible (2), but so far has not been used to a large extent, mainly because so few specific structures have been identified on the protoplast cell surface.

Because of the large size of cells and the need to keep them viable and capable of further growth, isolating whole viable cells presents a different problem from isolating soluble macromolecules. Thus, pH, salt concentration, temperature, nutrients, antibiotics, etc. are important factors, and it is necessary to work under sterile conditions. The density of cell surface molecules and their affinity for the chosen ligand are also critical.

1.2 Affinity separation of multi-protein complexes

A similar problem is present when cellular multi-protein complexes are isolated by affinity methods. Such experiments are usually performed in order to find unknown proteins which bind to the defined protein for which the affinity ligand is specific. Thus, a method must be found which breaks open the cells without disrupting the association between the proteins in the complex. The same types of matrix materials popular for affinity separations of macromolecules (see Chapter 1) are used. Many rather complicated methods have been worked out for isolating multi-protein complexes; for example, receptor kinases bind to many proteins via their Src homology (SH2 and SH3) domains and similar structures. It is very hard to give generalized protocols for these applications, as almost every individual protein–protein interaction requires its own method and optimization. These methods are, moreover, seldom quantitative or presented in quantitative terms.

The pace of development in this field is rapid, and it may change dramatically following the recent introduction of a powerful new genetic method, namely the two-hybrid system developed in yeast (3). The system is based on transcription of a gene necessary for survival in a selective medium. A transcription factor gene is engineered so that the one part is fused to a gene encoding the protein of interest, while a cDNA library or a mutated gene is fused to the other part. The transcription factor is active only when its two domains are juxtaposed by affinity between the two introduced fusion partners. The system can be used for finding unknown affinity ligands and for analysis of affinity.

Protocol 1. Detachment of monolayer cells prior to affinity separation[a]

Equipment and reagents

- PBSA (phosphate-buffered saline without calcium and magnesium, Dulbecco's phosphate buffered saline, solution A: 0.15 M NaCl, 1.9 mM NaH_2PO_4, 8.1 mM Na_2HPO_4, pH 7.4)
- PBS[b] supplemented with calcium and magnesium (Dulbecco's complete PBS, add $CaCl_2 \cdot 2H_2O$ at 132.5 mg/litre and $MgCl_2 \cdot 6H_2O$ at 100 mg/litre)
- Dissociation solution, 7 mM EDTA in PBSA
- Tissue culture medium supplemented with serum, chosen according to the growth and viability requirements for the cells used
- Incubator for tissue culture (CO_2, 37 °C)
- Standard inverted microscope
- Shaker/agitator
- laminar flow hood

Method

1. Grow the cells in conditions appropriate for the cell type, i.e. in the usual growth medium and with the usual supplements, in plastic bottles or Petri dishes. Split the cells at confluency, unless they are

Protocol 1. *Continued*

difficult to dissociate, in which case they should be grown to a lower density.

2. Remove the tissue culture medium and rinse the cells three times with PBSA (all solutions should be pre-warmed to 37°C) using the same volume as that of the medium removed. This step is important, as too much calcium remaining from the growth medium will interfere with the ability of EDTA to disrupt binding of cell surface structures with other cells or extracellular matrix proteins.
3. Add dissociation solution (use a volume equal to the volume of medium removed), and decant immediately, leaving just a thin layer of fluid on top of the cells. Incubate, agitating frequently, and monitor the cells under the microscope. The incubation time required for dissociation can vary depending on cell type. Typically it is 5–10 min, but it may take up to 2 h. If the incubation time is too long, viability will be affected. The optimum incubation time should be determined empirically for each individual cell line.
4. Tap the culture vessel cautiously with the fingers. When the cells are loose, add medium, or PBS supplemented with Ca^{2+} ions[c]. Use a Pasteur pipette to remove all cells. The cells[d] are now ready for affinity separation.

[a]With this method, anchorage-dependent cells can be detached without using proteolytic enzymes, which may remove antigens from the cell surface.
[b]PBS is available both as powder or as tablets from all major sources of tissue culture media.
[c]Note that Ca^{2+}-free conditions will affect viability. If the cells are to be recultured, they should be replenished with calcium after separation.
[d]All manipulations with cells in this chapter should be done under sterile conditions.

2. Affinity separation strategies for cells

Two main strategies have been developed for isolating single cell types from complex mixtures:

- *positive selection*, in which cells of a certain phenotype are tagged with a specific reagent, which makes it possible physically to remove them from the original population of cells and recover them for further work;
- *negative selection* (or depletion), in which the required cell type is enriched by removing unwanted cell types from the original cell mixture.

The choice of strategy is determined by factors such as the proportions of different cell types, the feasibility of removing specific reagents used for separation, and the inertness of the reagents used (some reagents can alter the viability of cells or interfere with the biological parameters under study). It is therefore recommended that, where possible, several separation methods are compared in order to select the one with the fewest adverse effects on the cells used.

The success of negative selection is entirely based on the efficiency of the removal of the contaminating cells, and thus on the availability of good affinity ligands for their removal. The method is most successful when used in a mixture of cells where the required cells make up a relatively high proportion of the population and the contaminating cell types are easy to monitor. If the required cell type occurs in a complex cell mixture, negative selection might be difficult, as it requires that the different contaminating cell types have to be removed one by one.

The major drawback of negative selection is the possibility of contamination of the selected population of cells by a low frequency of cells which are biologically active. An example of this is separation of blood cells from peripheral blood or spleen. It has been calculated that the admixture of only a few macrophages to separated B lymphocytes would affect their function (4).

In spite of its drawbacks and although it is more time-consuming than positive selection, many experimentalists prefer negative selection to positive selection at least when it comes to working with cells from peripheral blood or from the immune system. This is because many of the affinity ligands used in positive selection are not biologically inert and so may affect the properties of the selected cell. For instance, antibodies to the cell surface protein CD19 are efficient for positive selection of B-cells, but they are known both to activate and to inhibit B-cells (5,6).

Methods based on negative selection do not usually have such side effects, but exceptions have been reported. Thus, to enrich for B-cells, T-cells can be removed based on the affinity of the cell surface antigen CD2 to bind to sheep erythrocytes or using matrix bound anti-CD3 antibodies. However, this leads to activation of the T-cells (7,8), resulting in release of cytokines and potentially to stimulation of B-cells. Therefore, as an important rule of thumb, the isolated cell type should always be tested to check that the separation method does not interfere with the properties under study.

3. Separation methods

The methods described in this section can be used irrespective of whether the strategy used is positive or negative selection.

3.1 Cell adhesion to plastic surfaces

Cells from bone marrow or peripheral blood are often the starting material for affinity purification. The resulting purified cell populations are frequently contaminated by cells of the macrophage/monocyte lineage. Most researchers working with other cell types ignore the possible effects of macrophage contamination, which is unfortunate, as these cells are very active and may produce several lymphokines.

Macrophage/monocyte cells adhere to plastic surfaces and, although not based on specific affinity, this property can be exploited to remove them. The method will not work, of course, if the required cell type adheres too. In this method, cells are allowed to sediment in the culture dish for 1 h at 37 °C. Do not use too high a cell density (i.e $<10^5$ cells/cm^2). The non-adherent cells are gently removed together with the culture medium. The procedure should be repeated twice. Other methods for removing macrophages also exist (see Section 4.)

3.2 Affinity ligands

The use of specific ligands is a most attractive method for separating whole cells or multi-protein complexes. In the majority of cases polyclonal or monoclonal antibodies are used. The most commonly used reagents available have been developed for different types of haematopoietic cells, as they are directed against cell surface molecules, defined by sets of monoclonal antibodies recognizing well identified molecules, the so-called CD clusters (9). However, there is a growing list of examples in which cells from other tissues have been isolated by different affinity-based methods. *Table 1* shows a selection of some published examples. More information can usually be obtained from the suppliers of reagents.

Antibodies can be coupled to the separation matrix by a number of methods (see Chapter 6). Several commercially available, pre-coupled matrices exist, in which specific antibodies are coupled for isolation of specific cell types, such as T-cell subsets or B-cells. Of more general use are matrices pre-coated with immunoglobulin-binding molecules. Of these, staphylococcal Protein A and streptococcal Protein G, which bind to the Fc part of immunoglobulin molecules, have found a widespread use. Species-specific antibodies are used to coat the separation matrices to capture cells coated with antibodies from a range of species.

Compared with antibodies, adhesion molecules have in general a rather low affinity for their ligands. Thus, intercellular adhesion molecule-1 (ICAM-1 or CD54) has been found to have an affinity for its ligand, the integrin lymphocyte function-associated antigen-1 (LFA-1 or CD18/CD11a), in the micromolar range (10). However, matrix-bound ICAM-1 was found to be efficient for the binding of LFA-1-positive cells (11).

When specific antibodies or ligands are not available or if the affinity ligand cannot be bound to the matrix, other strategies have been developed. One of the most efficient is the biotin–avidin system. Biotin is a vitamin with an extraordinarily strong binding to avidin or streptavidin (K_d = 10^{-15} M). Biotin-labelled ligands can be captured by many commericially available avidin- or streptavidin-coupled matrices. Thus, following biotinylation of any molecule binding to a given cell, these cells can be captured by streptavidin-coupled matrices. Several kits are available for biotinylation from many commercial sources. To avoid steric hindrance caused when several biotinylated molecules bind to one avidin molecule, biotin is available with a choice of chemical spacers (e.g. from Pierce). Several other derivatives of biotin are also available. One practical way to biotinylate antibodies and other proteins is shown in *Protocol 2*.

Protocol 2. Biotinylation of antibodies and other proteins

Equipment and reagents

- Dialysis tubing
- 0.2 M borate buffer, pH 8.5 (Note: buffers containing amines cannot be used)
- Eppendorf tubes
- PBS (see *Protocol 1*)
- *N*-Hydroxysuccinimido-biotin (e.g. Sigma, Pierce)
- Dimethylsulfoxide (DMSO, Sigma)
- Agitator

Method

1. Dialyse 1–2 ml of antibody or other protein (of known concentration) against 0.5 litres of borate buffer at 4°C . Change the buffer after 2 h, replacing it with the same volume, dialyse overnight, and then transfer the sample to a Eppendorf tube.
2. Dissolve the biotin in DMSO to a final concentration of 1 mg/ml. The biotin should be prepared fresh and used immediately.
3. Add 100 μl of biotin solution per mg of antibody or protein.
4. Agitate for 4 h at room temperature.
5. To remove free biotin, dialyse against 0.5 litres of PBS at 4°C for 2 h. Replace buffer with 1 litre of fresh PBS and continue dialysis overnight. The antibody or protein can be supplemented with 0.1% (w/v) sodium azide to prevent microbial growth[a].

[a]The biotinylated protein is stable. It should be tested to exclude the possibility that the binding properties have changed relative to the original protein.

Using the biotin–streptavidin system it has been possible to exploit unconventional ligands for affinity separations, especially with experimentally modified cells. In a receptor-cloning approach Harada *et al.* (10) used biotinylated interleukin-4 (IL-4) to select cells that expressed the IL-4 receptor. The cells were selectively bound to plastic surfaces coated with anti-biotin antibodies by a process called panning (see *Protocol 3*). Intestinal cell lines with receptors for heat-stable enterotoxin were isolated using biotinylated toxin (13).

3.3 Separation methods

In most affinity-based separation methods the specific ligand is coupled to a solid phase, which allows easy and efficient physical separation of the cells of interest. The coupling may be either by covalent binding to the matrix or, as mentioned above, to proteins which then bind to the affinity ligand, usually antibodies.

Many sophisticated methods have been used for a long time, mostly in immunological studies. These include affinity column chromatography as well as cell sorting. These methods are efficient and specific, but they are also tedious and have a certain risk of microbial contamination. Fluorescence-activated cell sorting (FACS) methods—besides the need for sophisticated instruments—take a long time and are not practical for the isolation of large numbers of cells for biochemical or molecular studies. One unique advantage, though, is that sorting can be based on several parameters at the same time. Another feature worth considering is that when the purpose of the experiment is analytical, FACS analysis might obviate the need for cell separation. While cell sorting has been considered excellent for isolation of rare cells, more straightforward methods based on antibody-coated plastic surfaces or magnetic beads can be just as efficient.

For practical purposes two convenient methods have become predominant as they are simple, give high yields, and have little contamination from unwanted cells. One of these methods is the adhesion of cells to antibody-coated plastic surfaces, often referred to as panning. The other method is the use of magnetic beads, to which the specific ligand is coupled. Using very strong magnets the cell/magnetic bead mixture is separated from cells not bound to the ligand. These two methods are discussed further in Sections 3.3.1 and 3.3.2 respectively.

Other methods using matrix-bound antibodies or ligands are based on standard chromatographic procedures. Matrices like polyacrylamide, dextran, polystyrene, etc. are used; however, the particle size is larger than those used for biochemical separations to allow whole cells to pass through. The affinity system used can be biotin–avidin (Cellpro Inc.), Protein A (Pharmacia), or many others.

3.3.1 Panning

Adhesion to antibody-coated plastic surfaces, or 'panning', is a most practical method which was devised in 1978 (14). The method is described in a generalized form in *Protocol 3.* This method is excellent for positive selection, but is also often used for negative selection. It can be done under sterile conditions, is rapid, and usually gives a good yield. It does not require any special equipment and large number of cells can be processed.

i. Practical advice

One major disadvantage of panning is that sensitivity and cell yield can be low. Optimization by titration of antibody concentration, testing of optimal cell number in the cell mixture added to the plates, and varying the number of rinses in negative selection is advisable. When used in positive selection, the binding to the antibody could be very strong. This problem could be overcome by reducing antibody concentration or by mixing the specific antibody with a non-specific antibody. Blocking with BSA is important to avoid non-specific binding to the plastic.

Protocol 3. Isolation of specific cells by panning

Equipment and reagents

- Affinity ligand, usually monoclonal or polyclonal antibodies
- 50 mM Tris–HCl buffer, pH 9.6
- Petri dishes, bacteriological or tissue culture quality[a], 60 × 15 mm or 90 × 15 mm (Nunc, Falcon, or equivalent)[b]
- Phosphate-buffered saline (PBS) with 1 mg/ml bovine serum albumin (BSA)
- Tissue culture medium supplemented with 5% (v/v) serum
- Rubber policeman/cell scraper (Nunc, Costar, or equivalent)
- Bench-top centrifuge with microtitre plate holder, which is a standard attachment to most centrifuges (e.g. Beckman TJ-6R or equivalent)

Method

1. Dilute antibodies in 50 mM Tris–HCl, pH 9.6, to give a final concentration of 10 μg/ml.
2. Coat bacteriogical Petri dishes or tissue culture flasks with the antibody solution, 4 ml in the 90 × 15 mm dishes and 3 ml in the 60 × 15 mm dishes, and incubate at room temperature for 4 h or overnight[b].
3. Remove antibody solution and wash three times in PBS.
4. Block with PBS containing BSA for 15 min at room temperature.
5. Add the concentrated cell solution in tissue culture medium with 5% (v/v) serum and incubate at room temperature for 60 min. A typical cell number could be 10^5 cells/cm^2, higher if the desired cell population is rare.
6. Collection of cells:
   (a) Positive selection: Remove the supernatant, and discard. Rinse twice in PBS, in the same volume as the coating solution, and discard the supernatant. Remove the cells gently using a rubber policeman/cell scraper or by repeated pipetting of tissue culture medium.
   (b) Negative selection: Collect the supernatant and discard the adhering cells. Rinse the plates/bottle five times with PBS using the same volume as the coating solution, and add the rinsing solutions to the original supernatant with careful monitoring of contamination and cell yield.

[a]In the original description sterile polystyrene dishes of bacteriological quality were recommended, but tissue culture quality can be used, especially when complemented with a centrifugation step. The antibodies bind to the plastic surface without any need for coupling reactions.
[b]Alternatively, 25 cm^2 tissue culture plastic bottles can be used, which should be coated with 12 ml of antibody solution. After the addition of cells they can be centrifuged in a microtitre plate holder for 10 min at 200*g*, followed by another incubation for 50 min.

It is important that the antibody used for coating is pure and preferably monoclonal. Polyclonal antibodies (in the form of a purified IgG fraction) may be used, but have lower capacity because the numbers of IgG-binding sites on the plastic surfaces are fixed and saturable.

3.3.2 Immunomagnetic beads

The introduction of magnetic beads has been a major breakthrough for affinity-based methods for cell separation and this has become the predominant method for the separation of cell suspensions, especially for haematopoietic cells. The determining factor for success is the specificity of the antibody or of other affinity ligands used. If the antigen is expressed on more than the desired cell type, separation with immunomagnetic beads could be combined with other methods, such as centrifugation, growth cycles in selective media, etc.

There are several suppliers of superparamagnetic particles (i.e. ones that are only magnetic in a magnetic field). The properties of homogenous superparamagnetic beads, matrices, and their use have been reviewed (15,16). Different types of beads are available: the most widely used are Dynabeads (Dynal AS), polystyrene beads with a homogenous size and a diameter of 4.5 μm, and MACS Microbeads (Miltenyi Biotech GmbH), irregular microparticles based on modified dextran with a size ranging from 30 to 60 nm. The properties of the two systems are compared in *Table 2.* The size and the strong binding of the Dynabeads to the cells could in some cases interfere with functions and other properties of the selected cells and therefore they are more favoured for negative selection. The microparticles of the MACS Microbead system are much smaller and are biodegradable and theoretically lack these disadvantages. They are useful for both positive and negative selection. However, if cells shed the antigen used for selection, or if the cells

Table 2. Comparison of superparamagnetic beads from different sources

Property	Dynabeads	MACS Microbeads
Size	4.5 μm, homogeneous	30–60 nm, irregular
Matrix	polystyrene	modified dextran
Biodegradability	not biodegradable	biodegradable in the lysosomal compartment
Detachability	not usually detachable	not necessary, as they are small and can be degraded by the cells
Interference with biological properties	do not interfere, but bound antibody may	do not interfere, but bound antibody may
Interference with optical properties	could disturb FACS applications	does not disturb FACS applications
Best application	negative and positive selection	positive and negative selection

could be detached from the antibody (see *Protocol 7*) the two systems would be comparable. The Dynal system is more flexible as it can be adapted to Eppendorf tubes, microtitre plates, and other ordinary laboratory equipment. The process can be monitored by the naked eye. The MACS Microbead system, on the other hand, relies on special separation columns.

Other magnetic beads include IObeads (Immunotech), which are irregular, 1 μm microparticles, and Advanced Magnetics (Perseptive Diagnostic), which are 0.5 μm in diameter, where the antibodies are coupled by glutaraldehyde. They are similar in properties to the Dynal system; however, separation time increases as the size of the beads decreases. The Advanced Magnetics beads take up to four times longer than the Dynal beads to be removed by the magnet. On the other hand, these beads are much cheaper.

Two basic techniques can be applied with magnetic beads (as well as with other affinity separation systems). In the *direct technique* (as demonstrated in *Protocols 3B* and *5*), the magnetic particles are coated with the monoclonal antibody or other affinity ligand used before the cells are added. In the *indirect technique* the cells are pre-coated with the antibody/affinity ligand, before addition to the beads.

Protocol 4 describes an easy and convenient method for the separation of the starting cell population before the affinity separation step, here exemplified with human peripheral blood cells. *Protocol 4B* and *C* describe the separation of primary B-cells using Dynabeads with the direct and indirect methods respectively. *Protocol 5* demonstrates separation of mouse spleen cells using MACS Microbeads.

Protocol 4. Affinity separation of human primary B cells from peripheral blood using Dynabeads

Equipment and reagents

- Buffy coat cells[a]
- 50 ml sterile test tubes, tissue culture quality (Nunc, Falcon, or equivalent)
- Lymphoprep (Nycomed AS) or equivalent
- Physiological saline
- PBS
- Hank's balanced salt solution (BSS) with 1% fetal calf serum (FCS)
- Bench-top laboratory centrifuge, with swing-out rotor (e.g. Beckman TJ-6R or equivalent)
- Tissue culture medium, RPMI 1640 serum-free or with 1 or 10% (v/v) FCS
- Immunomagnetic beads coupled with CD19 antibodies (Dynabeads M-450 Pan-B) or sheep anti-mouse IgG
- Anti-CD19 monoclonal antibodies (Dako AS or equivalent)
- Magnet (MPC-1, Dynal AS) or equivalent
- 25 cm^2 tissue culture bottle (Nunc, Falcon, or equivalent)
- Tissue culture CO_2 incubator
- Haemocytometer or electronic particle counter (Coulter or equivalent)
- Laminar flow hood[b]

A. *Preparation of white blood cells*

Note: All steps should be done in a laminar flow hood; always use gloves.

1. Dilute one volume of buffy coat cells with at least two volumes of saline at room temperature.

Protocol 4. *Continued*

2. Add 37 ml of this suspension to a 50 ml sterile plastic tube and layer 10 ml of Lymphoprep below the diluted cell suspension.
3. Spin at 900*g* for 25 min at 18–20 °C with no brake.
4. Remove the interface, which contains mononuclear cells. Avoid too much admixture of the Lymphoprep phase.
5. Distribute cells from one buffy coat into three 50 ml tubes, dilute up to 45 ml with PBS and spin at 400*g* for 10 min.
6. Discard supernatant and wash twice in PBS at 400*g* for 10 min.
7. Dilute in Hank's BSS with 1% (v/v) FCS and count the cells in a haemocytometer or an electronic particle counter.

B. *Immunomagnetic separation of human primary B-cells by the direct method*[c]

1. Prepare the beads (4×10^8 beads/ml, 0.5 ml beads will be sufficient for the number of B-cells typically found in one buffy coat) by washing twice in Hank's BSS to remove NaN_3 added as bacteriostatic agent by the producer. Calculate a bead:cell ratio of 10:1.
2. Mix cells in medium with beads. It is convenient that cells from one buffy coat are diluted in 8 ml of medium (see step 7, *Protocol 4A*).
3. Incubate at 4 °C for 30 min with gentle tilting or rotating[d].
4. Place the magnet at the side of the test tube for 2–3 min and drain the supernatant, while the cells are kept on the wall by the magnet.
5. Remove the magnet and suspend the cells in 8 ml of Hank's BSS, 1% (v/v) FCS; repeat steps 4 and 5 four times.
6. Suspend the cells in 12 ml of medium with 1% (v/v) FCS, in a 25 cm^2 culture bottle and incubate overnight at 37 °C.
7. Separate detached beads from cells by passing the cell suspension through the magnetic field twice, 2–3 min each time. To avoid non-specific adsorption to the beads, the cell/bead mixture should be carefully pipetted.
8. Suspend the cells in growth medium and count.[e]

C. *Immunomagnetic separation of human primary B-cells by the indirect method*[f,g]

1. Incubate the cell suspension with the CD19 antibody for 30 min at 4 °C It is advisable to titrate each individual antibody used in this application to reduce costs. 0.5 $\mu g/10^6$ cells in a volume of 100 μl is usually sufficient.

2. Collect the pre-treated cells by centrifugation at 200*g* for 10 min and discard the supernatant.
3. Resuspend and wash the pre-treated cells twice with Hank's BSS pH 7.4.
4. The target cells treated with the primary mouse monoclonal antibody can now be isolated with the sheep anti-mouse IgG-coated beads according to *Protocol 3B.*

[a]These represent white cells obtained after fractionation of donated blood into red blood cells and plasma, which is a standard procedure in most blood banks. The buffy coat fraction is a by-product of no immediate value and is usually discarded. Different blood banks use different speeds for the centrifuation step, which could mean variable amounts of platelets. One transfusion unit (450 ml of blood) consists of 40–60 ml of white cells in plasma mixed with red cells. One buffy coat may contain $1–2 \times 10^7$ B-cells, but occasionally higher levels. Due to the risk of contamination with hepatitis viruses, HIV, and other infectious agents, it is not recommended that sources of human peripheral blood are used other than that from established blood banks with good routines for monitoring of infectious agents
[b]Horizontal flow hoods should not be used when working with blood cells, as blood should always be considered as a potential risk for infection and should be treated as such. As a rule vertical laminar flow hoods should be used, where the air does not blow against the investigator.
[c]The principle builds on the fact that CD19 is a B-cell-specific antigen expressed on all peripheral blood B-cells. The anti-CD19 antibody is directly coupled to the magnetic beads by the supplier.
[d]It is important that cells and solutions are kept at 4°C during steps 3–5, in order to prevent binding of macrophages or other adherent cells.
[e]B-cells will be detached from the anti-CD19 antibodies by spontaneous shedding of the antigen. This does not occur with most antibodies. *Protocol 7* shows a method for detaching cells from beads.
[f]In this example the same principle is followed as in the direct method. However, the CD19 antibodies are not directly coupled to the magnetic beads, but are first bound to the B-cells in the cell mixture, which is then added to the magnetic beads, to which sheep anti-mouse IgG antibodies are bound.
[g]*Protocol 4C* is an example to show the principle of the indirect method. However, the direct method is recommended for isolating B-cells, as coating with anti-CD19 antibodies may cause shedding of the CD19 antigen.

The indirect technique is more time-consuming, requires more antibody and must be followed by several washing steps. However, it is more efficent as the yield of the desired cells is usually higher. In most cases it is recommended that the indirect technique is used, as it is more efficient.

A frequently used application of the indirect technique is negative selection of a desired cell population. The method is very attractive, as the magnetic beads will not stick to the desired cell population as after positive selection. Nor will antibodies interfere with cell surface receptors, which might interfere with biological functions under study. Negative selection using the indirect technique should therefore always be considered. However, in practice antibodies towards all the possible contaminants are usually not available.

Protocol 5. Separation of mouse T-cells from spleen with the MACS Microbead system

Equipment and reagents

- Mouse spleen freshly removed after killing by cervical dislocation
- RPMI 1640 supplemented with 10% FCS and 50 μM 2-mercaptoethanol
- 90 × 15 mm Petri dish
- Scissors and forceps, kept sterile in 70% (v/v) ethanol
- 200-μm mesh stainless steel screen and a teaspoon, autoclaved
- 10 ml polystyrene centrifuge tube with conical bottom (Nunc, Falcon, or equivalent)
- Bench-top centrifuge
- MiniMACS starting kit with a separation unit, one stand, and sterile prepacked columns type MS, 1 ml (Miltenyi Biotec GmbH, art. no. 421-01)
- Thy 1.2 Microbeads (Milteneyi Biotech GmbH, art. no. 491-01) (only for mouse strains expressing Thy 1.2)
- PBS supplemented with 5 mM EDTA and 0.5 % (w/v) BSA
- Haemocytometer or an electronic particle counter

A. *Preparation of spleen white cells*

1. Place the spleen in the Petri dish together with a few millilitres of medium; using the scissors cut it into small pieces.
2. Place the screen in another Petri dish, and press the spleen fragments through the screen with the spoon using gentle circular movements.
3. Collect material which has passed through in a 10 ml tube. Rinse the Petri dish with medium until the tube is filled up. Allow larger fragments to sediment for 3 min and transfer the supernatant to a new tube.
4. Wash the cells in 10 ml of medium by centrifuging at 400*g* for 10 min and resuspend the cells in 5 ml of medium. Count the cells in a haemocytometer or an electronic particle counter.

B. *Separation of T-cells*

1. Centrifuge cells at 400*g* for 10 min and resuspend cell pellet in 90 μl of supplemented PBS per 10^7 cells.
2. Add 10 μl of Thy 1.2 Microbeads/10^7 cells and incubate at 4°C for 15 min.
3. Place the separation column in the separation unit.
4. Fill column with 500 μl of supplemented PBS, which will take <3 min to flow through. Discard the effluent. The column is a stop-flow column which will not run dry.
5. Place the labelled cells on top of the column, allow the suspension to pass through, and wash with 500 μl of EDTA-supplemented PBS.
6. Wash twice with 500 μl of supplemented PBS.

7. Remove column from the separator, add 1 ml of supplemented PBS, and flush out the magnetically labelled cells into a suitable tube with the plunger supplied with the column.

The method is adapted for 10^4–10^7 positive cells, but could be scaled up to 10^9 using other equipment available from the same supplier.

i. Practical advice

The density of antibodies or antibody-binding molecules on immunomagnetic beads has been adjusted by commercial suppliers to give a binding capacity suitable for practical application. Affinity constants are difficult to measure on the beads and are seldom given by the suppliers, rather a bead:cell ratio is recommended. In general terms, with a high bead:cell ratio, the full capacity of the beads is not used, and with a low ratio there is a loss of some of the cells of interest. Usually, the rate of contamination with irrelevant cells is higher in the former case. A simple rule is that in positive selection, the bead:cell ratio should be lower than in negative selection.

The situation is further complicated if the fraction of required cells in a cell mixture is small or variable. This is especially important for tumour cells, where the density of antigen can deviate considerably from what is found in the corresponding normal cells. For economic reasons or, more specifically, to combine high capacity with high purity, the bead:cell ratio should be optimized as well as the antibody concentration for each type of application.This can be done in two steps:

1. Titrate the amount of antibody (or affinity ligand) which is to bind to the matrix with fixed concentrations of cells and beads. *Protocol 4* can be used.
2. Test different bead:cell ratios, e.g. from 20:1 to 1:1. Cell numbers recovered and their purity should be measured with methods which are relevant for the required cell type.

There are several practical tools available commercially to make the immunomagnetic separation easy and convenient. Holders for magnets, magnets suitable for microtitre wells, and magnets and other tools for large-scale separations are available. In the Dynal system Tosyl-activated beads for direct coupling are available.

The use of immunomagnetic beads is well established for haematopoietic cells. When used with other cell types, it is advisable to check for endocytosis of the beads, which has been reported to occur in myogenic cells (17).

3.3.3 Immunomagnetic separation of cells using ligands with low affinity

If the ligand used has low affinity, the indirect methods as shown in *Protocols 4* and *5* will not work. This is often the case with non-antibody ligands like adhesion molecules. A modification of the indirect method is shown in *Protocol 6*. The purpose of this is to isolate cells that express the integrin LFA-1 in its active conformation. This molecule is expressed on many haematopoietic cells with low affinity, but affinity for its ligand ICAM-1 is increased when the cells are activated at 37°C (18). The protocol can easily be modified for other fusion proteins.

In the protocol outlined a fusion protein is used in which the specifically binding part is the extracellular domain of ICAM-1 and the C-terminal part is the Fc-domain of a mouse IgG_{2b} immunoglobulin (19). The standard protocol for indirect method will not work, as the affinity for ICAM-1 is in the micromolar range even on activated cells (10) and therefore coating of the cells with the fusion protein will be too inefficient. If the *beads* are coated instead of the cells, the density of the ligand will be high enough for the cells to bind to the beads. The same procedure can be used for isolating very rare mutagenized cells, if separation and growth are done repeatedly.

Protocol 6. Affinity separation with magnetic beads using a ligand with low affinity

Equipment and reagents

- Cell lines expressing the fusion protein, here exemplified by the ICAM-1/mouse IgG_{2b} Fc domain, which was produced in CHO cells (9)
- The fusion protein, affinity-purifed on a Protein A column (Pharmacia), can be used but, for most practical applications, the tissue culture fluid can be used directly, adjusted to 20 μg/ml without serum, but supplemented with BSA at 1 mg/ml
- Immunomagnetic beads coupled with sheep anti-mouse IgG (Dynabeads M-450, product no 110-03).
- BSA, 1 mg/ml in PBS
- Mouse IgG, 100 μg/ml in PBSA
- Magnet (MPC-1, Dynal or equivalent)
- Shaker/rotater

Method

1. Wash 1 ml of anti-mouse IgG beads with PBSA/BSA twice.
2. Decant and add 1 ml of the solution containing the fusion protein.
3. Incubate for 24 h at 4°C with gentle tilting and rotation.
4. The beads can be stored for several months at 2–8°C in the coating solution until use under sterile conditions (0.1% w/v azide).
5. Before use, block free anti-immunoglobulin. Take a volume of beads appropriate to the volume of cells (in this case 20 beads/cell, but this should be optimized for each application) and remove the coating solution by placing the magnet at the side of the test tube for 2–3 min. Drain the supernatant, while keeping the beads on the wall using the magnet. Add the same volume of PBS supplemented with mouse IgG. Incubate for 1–2 h at room temperature with occasional gentle vortexing.
6. Wash twice and resuspend to the original volume.

7. Add the beads to the cells to be isolated in a ratio of 20:1; incubate with gentle tilting and rotation for 7 min.
8. Count cells[a].

[a]A convenient way to analyse LFA-1-dependent binding is to include as a control an anti-LFA-1 monoclonal antibody to show specificity. Chromium-51 labelling of the cells has been used for quantification (19).

3.3.4 Elution of cells after immunoaffinity separation

Antibodies are attractive for affinity separation of cells due to their high selectivity. By selection of monoclonal antibodies desired properties can be obtained, e.g. ability to retain the cells during rinsing or magnetic separation. However, after separation, in many cases the beads must be removed, to avoid interference with the experimental system.

In some cases the cell surface antigen is shed from the cells together with the used ligand by incubation overnight at 4 °C. As mentioned above this is the case for CD19 antibodies, which are used for the positive selection of primary B-lymphocytes.

A more rapid method has been developed where an anti-Fab polyclonal antibody (20) interacts with the coupled antibody and releases some but not all bead-labelled antibodies. The procedure is outlined in *Protocol 7*.

When adhesion proteins are used as ligands, the cells may be released by metal chelators. This principle is shown in *Protocol 5*.

Protocol 7. Release of immunomagnetic beads from separated cells[a]

Equipment and reagents

- Goat anti-mouse Fab antibodies (DETACH-aBEAD, Dynal AS)[b]
- Other reagents as for *Protocol 3*

Method

1. Resuspend the cells (*Protocol 4B*, after step 5) in 100 μl of tissue culture medium with 10% (v/v) serum.
2. Add 10 μl of anti-Fab antibodies to the bead/cell conjugates.
3. Incubate at room temperature for 60 min with gentle tilting and rotation. Higher or lower temperature will reduce the yield of released cells.
4. Remove the detached beads as in *Protocol 4B*, step 7.

[a]This protocol is adjusted for separation of up to 10^7 cells bound to immunomagnetic beads in the ratio of 5–10 beads/cell (see *Protocol 4*).
[b]The supplier emphasizes that the system only works for certain antibody-coated beads.

Elution methods suffer from the risk of denaturing cell surface proteins, or may have other adverse affects on living cells. Physical separation methods require a reasonably high affinity binding in order to be able to remove the desired cells from the cell mixture. To break the binding may therefore require rather harsh methods. As discussed in Section 2, negative selection is a way to avoid this problem.

Besides the DETACHaBEAD anti-Fab antibody only one other system exists for removing immunomagnetic beads (21). It requires that DNA that can specifically bind a protein (the *lac* repressor) is coupled to beads. A fusion is made between the DNA-binding protein and Protein A. Cells bound via a monoclonal antibody to Protein A are separated and the beads are removed by DNase treatment, but the antibody and the fusion protein are left on the cell surface. The method is elegant, and conceptually important as it provides a means of breaking the strong interactions between beads and cells, which could be developed further.

4. Comparison of methods based on immunomagnetic beads with other separation methods

Separations based on panning or immunomagnetic beads are gentle methods for cell separation, and much more rapid than centrifugation methods, cell sorting in FACS machines, or classical chromatographic methods. A FACS instrument can sort around 10^7 particles per hour, while the same sorting with magnetic beads is complete after a few minutes. Large numbers of cells can be separated using very simple tools with a high degree of purity. Separation can be done in culture media or in salt solutions, which are well established for tissue culture.

If the antibody supply is limiting, panning is less cost-effective than using immunomagnetic beads, due to the geometry of the system. However, empirically it has been found to be just as effective for isolation of rare cells as magnetic beads. Seed and Aruffo (22) have devised a system for cloning receptor genes by repeated cycles of combined panning and growth of rare positive cells.

Affinity chromatography methods have been employed for cell separations for many applications. A typical example is the separation of T-cells using rabbit anti-mouse IgG antibodies coupled to large polyacrylamide beads (23). Isolated leukocytes were coated with different anti-T-cell-specific monoclonal antibodies and passed through a 5 ml column containing the beads. Another approach has been to use two-phase systems with PEG-modified antibodies (24). Two-phase systems have been used in elegant model systems for separating red cell populations (25). Judging from reported purities, efficiency, convenience, and time needed it is obvious why panning and immunomagnetic systems have become so popular in practical applications in recent years.

Adherence, or rather non-adherence, to nylon wool has often been used to remove monocytes and B-cells and to enrich for T-cells, which do not adhere to the same extent (26). The effluent cells contain T-cells in the range of 80–90% and both B-cells and monocytes are contaminants. This method has therefore often been used in the past as a first step in separation of T-cells, followed by more specific methods. A less complicated and less time-consuming method for decreasing contamination of monocytes is given in Section 3.1.

One of the most striking advantages of the new magnetic systems is that they obviate the need for tedious preparations of the starting cell populations. Although the protocols presented in this chapter start with the leukocyte fraction (*Protocol 4b*), the yield and purity are reasonable even when the starting material is whole blood.

For most practical cell separation applications the use of immunomagnetic beads has become the method of choice and it is safe to say that if a good antibody/affinity ligand exists the method is superior to any other method on the criteria of purity, cell viability, speed, and cost.

5. Monitoring purity

Cells which have been separated using affinity-based methods can be tested for purity by flow cytometry or immunofluorescence methods. Typically the number of contaminating cells is low, less than 1%. It is advisable when adapting the protocols for non-haematopoietic cells to optimize the recommended washing steps for yield and purity. Depending on the experimental system, the experimenter must evaluate the significance of the contamination.

If the purity is not sufficient we recommend that selection is continued using other panning or immunomagnetic bead steps. As long as there is an antibody available, these methods are more efficient and much milder than often-used methods based on complement-mediated lysis.

6. Examples of the application of affinity separation

Affinity separation of cells has found its most widespread use in the study of haematopoietic cells, especially immunocompetent cells. Separation of different stem cells in the bone marrow is another rapidly expanding field. For diagnostic purposes, methods have been established to separate fetal cells from the blood of pregnant women (27). Purging of bone marrow from tumour cells (28) or T-cells (29) before transplantation is another important application.

However, in many cases, affinity separation or, for that matter, any separation method does not need to be complete, as long as the method is mild enough to preserve the proliferative capacity and the desired cells have a growth advantage. This can be exemplified by the isolation of T-cells, which can be enriched from a complex mixture of blood cells by giving them the right growth conditions. A simple way of doing this is to isolate the white cell fraction according to *Protocol 3A*, and then activate them with anti-CD3 antibodies and add IL-2 to the culture (8). After 10 days all other cells are dead or have been overgrown. By different modifications the proportions of T-cell subsets can be modified or they can be directly sorted out, using FACS technology, or immunomagnetic bead separation protocols with, for example, anti-CD4 or anti-CD8 antibodies.

Isolation of B-cells is more controversial as both positive and negative selection have advantages and disadvantages. In negative selection protocols, adherent cells are removed by incubation on plastic surfaces, and T-cells are removed by immunomagnetic beads or other methods. In positive selection, B-cells are usually isolated by the expression of the CD19 antigen. Both methods give approximatively the same results with respect to cell yields and purity, although negative selection is much more time-consuming and requires several more steps. It has been argued that anti-CD19 antibodies can be either inhibitory or activating, as discussed in Section 2. However, CD19 has a rapid turnover, and cells separated in this way have been shown to be functionally active (29). It is advised that investigators check both approaches in the particular experimental system used, and that they do not just assume that the rapid and convenient positive selection methods are superior.

Transfection of DNA has become an attractive method for studying genes of interest. In practice, this is difficult, one reason being that transfection efficiencies are usually low, and the death rate high. (This is especially true for primary haematopoietic cells.) A protocol has been developed for studies of gene functions by transfection of primary B-cells. A truncated extracellular CD2 derivative was co-transfected together with the gene under study (31). The expressed CD2 could be isolated either by panning or by using immunomagnetic beads.

References

1. Rothenberg, E. V. (1991). *Methods*, **2,** 168.
2. Dörr, I., Miltenyi, S., Salamini, F., and Uhrig, H. (1994). *Bio/Technology*, **12**, 511.
3. Felds, S. and Song, O.-K. (1989). *Nature*, **340**, 245.
4. Howard, M., Mizel, S. B., Lachman, L., Ansel, J., Johnson, B., and Paul, W. E. (1983). *J. Exp. Med.*, **157**, 1529.
5. Carter, R. and Fearon, D. T. (1992). *Science*, **256**, 105.
6. Barrett, T. B., Shu, G. L., Draves, K. E., Pezzutto, A., and Clark, E. A. (1990). *Eur. J. Immunol.*, **20**, 1053.
7. Larsson, E.-L., Andersson, J., and Coutinho, A. (1978). *Eur. J. Immunol.*, **8**, 693.

8. Brattsand, G., Friedrich, B., and Gullberg, M. (1989). *Scand. J. Immunol.*, **30**, 233.
9. Schlossman, S. F., Boumsell, L., Gilks, W., Harlan, J. M., Kishimoto, T., Morimoto, C., *et al.* (1994). *Immunol Today*, **15**, 98.
10. Lollo, B. A., Chan, K. W. H., Hanson, E. M., Moy, V. T., and Brian, A.A. (1993). *J. Biol. Chem.*, **268**, 21693.
11. Hedman, H., Brändén, H., and Lundgren, E. (1992). *J. Immunol. Methods*, **146**, 203.
12. Harada, N., Castle, B. E., Gorman, D. M., Itoh, N., Schreurs, J., Barrett, R. L., Howard, M., and Miyajima, A. (1990). *Proc. Natl Acad. Sci. USA*, **87**, 857.
13. Almenoff, J. S., Williams, S. I., Schevin, L. A., Judd, A. K., and Schoolnik, G. K. (1993). *Mol. Microbiol.*, **8**, 865.
14. Wysocki, L. J. and Sato, V. L. (1978). *Proc. Natl Acad. Sci. USA*, **75**, 2844.
15. Haukanes, B.-I. and Kvam, C. (1993). *Bio/Technology*, **11**, 60.
16. Miltenyi, S., Muller, W., Weichel, W., and Radbruch, A. (1990). *Cytometry*, **11**, 231.
17. Prattis, S. M., Gebhart, D. H., Dickson, G., Watt, D. J., and Kornegay, J. N. (1993). *Exp. Cell Res.*, **208**, 453.
18. Dustin, M. L. and Springer, T. A. (1991). *Annu. Rev. Immunol.*, **9**, 27.
19. Hedman, H., Brändén, H., and Lundgren, E. (1992). *J. Immunol. Methods*, **146**, 203.
20. Rasmussen, A.-M., Smeland, E. B., Erikstein, B. K., Caignault, L., and Funderud, S. (1992). *J. Immunol. Methods*, **146**, 195.
21. Ljungquist, C., Lundeberg, J., Rasmussen, A.-M., Hornes, E., and Uhlén, M. (1993). *DNA Cell Biol.*, **12**, 191.
22. Seed, B. and Aruffo, A. (1987). *Proc. Natl Acad. Sci. USA*, **84**, 3365.
23. Braun, R., Teute, H., Kirchner, H., and Munk, K. (1982). *J. Immunol. Methods*, **54**, 251.
24. Karr, L. J., Donnelly, D. L., Kozlowski, A., and Harris, J. M. (1994). In *Methods in enzymology*, Vol 228 (ed. H. Walter and G. Johansson), p. 382. Academic Press, London.
25. Walter, H., Raymond, F. D., and Fisher, D. (1992). *J. Chromatogr.*, **609**, 219.
26. Julius, M. H., Simpson, E., and Herzenberg, L. A. (1973). *Eur. J. Immunol.*, **3**, 645.
27. Greifman-Holtzman, O., Blatman, R. N., and Bianchi, D. W. (1994). *Semin. Perinatol.*, **18**, 366.
28. Gribben, J. G., Freedman, A. S., Njeuberg, D., Roy, D. C., Blake, K. W., and Wood, S. (1991). *N. Engl. J. Med.*, **325**, 1525.
29. Koegler, G., Capdeville, A. B., Hauch, M., Bruester, H. T., Goebel, U., Wernet, P., and Burdach, S. (1990). *Bone Marrow Transplant.*, **6**, 163.
30. Funderud, S., Erikstein, B., Åsheim, H. C., Nustad, K., Stokke, T., Blomhoff, H. K., Holte, H., and Smeland, E. B. (1990). *Eur. J. Immunol.*, **20**, 201.
31. Pilon, M., Gullberg, M., and Lundgren, E. (1991). *J. Immunol.*, **146**, 1047.
32. Jamer, I. E., Walsh, S., Dodds, R. A., and Gowen, M. (1991). *J. Histochem. Cytochem.*, **39**, 905.
33. Camu, W. and Hendersson, C. E. (1992). *J. Neurosci. Methods*, **44**, 59.
34. Padmanabhan, R., Tsuruo, T., Kane, S. E., Willingham, M. C., Howard, B. H., Gottesman, M. M., and Pastan, I. (1991). *J. Natl Cancer Inst.*, **83**, 565.
35. Jackson, C. J., Garbett, P. K., Nissen, B., and Schrieber, L.(1990). *J. Cell Sci.*, **96**, 257.

36. Polanowska-Grabowska, R., Raha, S., and Gear, A. L. (1992). *Br. J. Haematol.*, **82**, 715.
37. Fujioka, T., Terasaki, P. I., and Heintz, R. (1990). *Transplantation*, **49**, 404.
38. Morris, J., Alaibac, M., Jia, M.-H., and Chu, T. (1992). *J. Invest. Dermatol.*, **99** 237.
39. Peter, A. T., Jones, P. P., and Robinson, J. P. (1993). *Theriogenology*, **40**, 1177.

A1

List of suppliers

Affinity Chromatography Ltd (ACL), 307 Huntingdon Road, Cambridge CB3 0JX, UK.

Affinity Sensors, Saxon Way, Bar Hill, Cambridge CB3 8SL, UK. 53 West Century Road, Paramus, NJ07652, USA.

Aldrich Chemical Co. Ltd, New Road, Gillingham, Kent SP8 4JL, UK.

Allied Colloids

Allied Colloids, 2301 Wilroy Road, PO Box 820, Suffolk VA 23439, USA.

Allied Colloids Ltd., Cleckheaton Road, Low Moor, Bradford W. Yorks. BD12 0JZ, UK.

Alltech Associates

Alltech Associates (UK), 6–7 Kellet Road Industrial Estate, Kellet Road, Carnforth, Lancs. LA5 9XP, UK

Alltech Associates 2051 Waukegan Road, Deerfield, IL 60015, USA.

Amersham

Amersham International plc., Lincoln Place, Green End, Aylesbury, Bucks. HP20 2TP, UK.

Amersham Corporation, 2636 South Clearbrook Drive, Arlington Heights, IL 60005, USA.

Amicon

Amicon Inc., 72 Cherry Hill Drive, Beverly, MA 01915, USA.

Amicon (UK), Upper Mill, Stonehouse, Gloucs. GL10 2BJ, UK.

Aminco, SLM Instruments Inc., American Instrument Co., Urbana Illinois 61801, USA.

Anatrace Inc.,1280 Dussel Drive, Maumee, OH 43537, USA.

Anderman

Anderman and Co. Ltd., 145 London Road, Kingston-Upon-Thames, Surrey KT17 7NH, UK.

J. T. Baker, Wyvols Court, Swallowfield, Reading, Berks. RG7 1PY, UK.

The Baker Co. Inc., PO Drawer, E. Sanford, Airport Road, Sanford, ME 4073, USA.

Bausch & Lomb

c/o *Spectronics Corporation*, 956 Brush Hollow Road, Westbury, NY 11590, USA.

c/o *LSI (UK) Ltd*, Unit 5, The Ringway Centre, Edison Road, Basingstoke, Hants. RG21 6YH, UK.

BDH (British Drug Houses) Ltd., Poole, Dorset, UK.

Beckman Instruments

Beckman Instruments UK Ltd., Oakley Court, Kingsmead Business Park, London Road, High Wycombe, Bucks. HP11 1J4, UK.

Beckman Instruments Inc., PO Box 3100, 2500 Harbor Boulevard, Fullerton, CA 92634, USA.

Becton Dickinson

Becton Dickinson and Co., Between Towns Road, Cowley, Oxford, OX4 3LY, UK.

Becton Dickinson and Co., 2 Bridgewater Lane, Lincoln Park, NJ 07035, USA.

Biocell Laboratories Inc., 2001 University Drive, Rancho Dominguez, CA 90220-6411, USA.

Bio

Bio 101 Inc., c/o Statech Scientific Ltd, 61–63 Dudley Street, Luton, Beds. LU2 0HP, UK.

Bio 101 Inc., PO Box 2284, La Jolla, CA 92038–2284, USA.

Bioprocessing

Bioprocessing Ltd., Medomsley Road, Consett, DH8 6TJ, UK.

Bioprocessing Inc.,116 Village Boulevard, Suite 200, Princeton, NJ 08540-5799. USA.

Bio-Rad Laboratories

Bio-Rad Laboratories Ltd., Bio-Rad House, Maylands Avenue, Hemel Hempstead, HP2 7TD, UK.

Bio-Rad Laboratories, Division Headquarters, 3300 Regatta Boulevard, Richmond, CA 94804, USA.

Biosepra

Biosepra (UK), Clarendon House, 125 Shenley Road, Borehamwood, Herts. WD6 1AG, UK

Biosepra, 140 Lock Drive, Marlborough, MA 1752, USA.

Boehringer Mannheim

Boehringer Mannheim UK (Diagnostics and Biochemicals) Ltd, Bell Lane, Lewes, East Sussex BN17 1LG, UK.

Boehringer Mannheim Corporation, Biochemical Products, 9115 Hague Road, P.O. Box 504 Indianapolis, IN 46250–0414, USA.

Boehringer Mannheim Biochemica, GmbH, Sandhofer Strasse 116, Postfach 310120 D-6800 Ma 31, Germany.

Cellpro Inc., 22215 26th Avenue, Bothwell, WA 98021, USA.

Clifmar Associates, School of Biological Sciences, University of Surrey, Guildford GU2 5XH, Surrey, UK.

Costar

Costar UK Ltd, Victoria House, 28–38 Desborough Street, High Wycombe, Bucks. HP11 2NF, UK.

Costar Corporation, One Alewife Center, Cambridge, MA 2140, USA.

Coulter

Coulter Scientific Instruments, Nothwell Drive, Luton, Beds. LU3 3RH, UK.

Coulter Corporation, PO Box 169015, Miami, FL 33116-9015, USA.

CPG Inc., 3 Borinski Road, Lincoln Park, NJ 7035, USA.

Dako Diagnostics, Denmark House, Angel Drove, Ely, Cambs. CB7 4ET, UK; 6392 Via Real, Carpenteria, CA 93013, USA.

Denley Instruments

Denley Instruments (Luckham Division), Victoria Gardens, Burgess Hill, West Sussex, RH15 9QN, UK.

Denley Instruments Ltd, Natts Lane, Billinghurst, Sussex RH14 9EY, UK.

Denley Instruments, PO Box 13958, Research Triangle Park, Durham, NC 22709-3958, USA.

Dextra Laboratories Ltd, The Innovation Centre, The University, PO Box 68, Reading, RG6 6BX, UK.

Difco

Difco Laboratories Ltd, PO Box 14B, Central Avenue, West Molesey, Surrey, KT8 2SE, UK.

Difco Laboratories, PO Box 331058, Detroit, MI 48232-7058, USA.

Du Pont

Du Pont (UK) Ltd, Industrial Products Division, Wedgwood Way, Stevenage, Herts. SG1 4Q, UK.

Du Pont Co. (Biotechnology Systems Division), PO Box 80024, Wilmington, DE 19880-002, USA.

Dynal

Dynal AS, 5 Delaware Drive, Lake Success, NY 11042, USA.

Dynal UK, Croft Business Park, 10 Thursby Road, Bromsborough, Wirral L62 3PW, UK.

European Collection of Animal Cell Culture, Division of Biologics, PHLS Centre for Applied Microbiology and Research, Porton Down, Salisbury, Wilts. SP4 0JG, UK.

Falcon (Falcon is a registered trademark of Becton Dickinson and Co.).

Fisher Scientific Co., 711 Forbest Avenue, Pittsburgh, PA 15219-4785, USA.

Fisher Scientific Equipment, Bishop Meadow Road, Loughborough, Leics. LE11 0RG, UK; 55 Cherry Hill Drive, Beverly, MA 1915, USA.

Fisons, *see* Fisher Scientific Equipment.

Flow Laboratories, Woodcock Hill, Harefield Road, Rickmansworth, Herts. WD3 1PQ, UK.

Fluka

Fluka-Chemie AG, CH-9470, Buchs, Switzerland.

Fluka Chemicals Ltd., The Old Brickyard, New Road, Gillingham, Dorset SP8 4JL, UK.

FMC

FMC Biosupport Materials, PO Box 20222, Pine Brook, NJ 07058-2022, USA.

FMC Bioproducts Europe, Risingenejl, Vallensback Strand, DK-2665, Denmark.

Gibco BRL

Gibco BRL (Life Technologies Ltd), Trident House, Renfrew Road, Paisley PA3 4EF, UK.

Gibco BRL (Life Technologies Inc), 3175 Staler Road, Grand Island, NY 14072–0068, USA.

Gilson

Gilson Inc., 3000 W. Beltline Highway, PO Box 62007, Middleton WI 53562-0027, USA.

c/o *Anachem Ltd*, Charles Street, Luton, Beds. LU2 0EB, UK.

GlycoTech, 14915 Broschart Road, Rockville, MD 20850, USA.

Hewlett Packard

Hewlett Packard Ltd, King Street Lane, Winnersh, Wokingham, Berks. RG11 5AR, UK.

Hewlett Packard, 3495 Deer Creek Road, Palo Alto, CA 94304, USA.

Arnold R. Horwell, 73 Maygrove Road, West Hampstead, London NW6 2BP, UK.

Hybaid

Hybaid Ltd, 111–113 Waldegrave Road, Teddington, Middlesex TW11 8LL, UK.

Hybaid, National Labnet Corporation, PO Box 841, Woodbridge, NJ 07095, USA.

HyClone Laboratories 1725 South HyClone Road, Logan, UT 84321, USA.

IBF, *see* BioSepra

ICN Biomedicals

ICN Biomedicals Ltd, Thame Business Park, Wenman Road, Thame, Oxon. OX9 3XA, UK.

ICN Biomedicals, 3300 Hyland Avenue, Costa Mesa, CA 92626, USA.

Immunotech International, BP 177, 130 Avenue de Lattre de Tassigny, Marseille, Cedex 913726, France; 160B Larrabee Road, Westbrook, ME4092, USA.

International Biotechnologies Inc., 25 Science Park, New Haven, CT 06535, USA.

Invitrogen Corporation

Invitrogen Corporation 3985 B Sorrenton Valley Building, San Diego, CA. 92121, USA.

Invitrogen Corporation c/o British Biotechnology Products Ltd., 4–10 The Quadrant, Barton Lane, Abingdon, OX14 3YS, UK.

Kem-en-Tec

Kem-en-Tec A/S, Lerso Parkalle 42.2, Copenhagen, DK-2100, Denmark.

(UK): *Biostat Ltd*, Pepper Road, Hazel Grove, Stockport, SK7 5BW, UK.

Kodak: Eastman Fine Chemicals 343 State Street, Rochester, NY, USA.

Lab m Ltd, Topley House, PO Box 19, Bury, Lancs. BL9 6AU, UK.

Lab Line Instruments Inc.,15th & Bloomingdale Avenue, Melrose Park, IL 60160-1491, USA.

Laboratory Data Control, 12431 Fairpoint Drive, Houston, TX 77099-3003, USA.

Life Technologies Inc., 8451 Helgerman Court, Gaithersburg, MN 20877, USA.

Luckham, *see* Denley Instruments.

3M

3M Corporation, 3M Center Building, 260-6B-16, St Paul, MN 55144-1000, USA.

3M Laboratories Ltd, 3M House, Morley Street, Loughborough, Leics. LE11 1EP, UK.

Merck

Merck Industries Inc., 5 Skyline Drive, Nawthorne, NY 10532, USA.

Merck, Frankfurter Strasse, 250, Postfach 4119, D-64293, Germany.

Millipore

Millipore (UK) Ltd., The Boulevard, Blackmoor Lane, Watford, Herts WD1 8YW, UK.

Millipore Corp./Biosearch, PO Box 255, 80 Ashby Road, Bedford, MA 01730, USA.

Miltenyi Biotech, 251 Auburn Ravine Road, Auburn, CA 95603, USA; Suite 4, Coliseum Business Centre, Riverside Way, Camberley, Surrey GU15 3YL, UK.

MSE Scientific Instruments, Manor Royal, Crawley, West Sussex RH10 2QQ, UK.

Nalge (Nunc), Foxwood Court, Rotherwas, Hereford, HR2 6JQ, UK; 75 Panorama Creek Drive, Box 20365, New York, NY 14602-0365, USA.

New England Biolabs (NBL)

New England Biolabs (NBL), 32 Tozer Road, Beverley, MA 01915–5510, USA.

New England Biolabs (NBL), c/o CP Labs Ltd, P.O. Box 22, Bishops Stortford, Herts CM23 3DH, UK.

Nikon Corporation, Fuji Building, 2–3 Marunouchi 3-chome, Chiyoda-ku, Tokyo, Japan.

Nunc, *see* Nalge.

Nycomed

Nycomed AS, Nycovejen 2, PO Box 42220, Torshov, N-0411, Oslo, Norway.

Nycomed (UK), Nycomed House, 2111 Coventry Road, Sheldon, Birmingham B26 3EA, UK.

Nygene Corporation, One Odell Plaza, Yonkers, NY 10701, USA.

Omnifit, 2 College Park, Coldhams Lane, Cambridge CB1 3HD, UK; 8 Executive Drive, PO Box 450, Toms River, NJ 08754, USA.

Oxford Glycosciences

Oxford Glycosciences Inc., Gross Island Plaza, 133 Brookville Boulevard, Rosedale, NY 11422, USA.

Oxford Glycosciences Ltd, Unit 4, Hitching Court, Blacklands Way, Abingdon, Oxon. OX14 1RG, UK.

Perkin-Elmer

Perkin-Elmer Ltd., Chalfont Road, Seer Green, Beaconsfield, Bucks, HP9 1QA, UK.

Perkin-Elmer-Cetus (The Perkin-Elmer Corporation), 761 Main Avenue, Norwalk, CT 0689, USA.

Perseptive Biosystems, 3 Harforde Court, Foxholes Business Park, Hertford SG13 7NW, UK; 500 Old Conecticut Path, Framingham, MA 01701, USA.

Perseptive Diagnostics, 735 Concord Avenue, Cambridge, MA 2138, USA.

Pharmacia Biosystems

Pharmacia Biosystems Ltd. Pharmacia Biotech, 23 Grosvenor Road, St. Albans, Herts. AL1 3AW, UK.

Pharmacia LKB Biotechnology AB, Björngatan 30, S-75182 Uppsala, Sweden.

Pharmacia Biotech Europe Procordia EuroCentre, Rue de la Fuseé 62, B-1130 Brussels, Belgium.

Pharmacia Biotech Inc., 800 Centennial Avenue, PO Box 1327, Piscataway, NJ 08855-1327, USA.

Phase Separations, Deeside Industrial Park, Deeside, Clwyd CH5 2NU, UK; 140 Water Street, Norwalk, CT 06854, USA.

Pierce, 3747 North Meridian Road, PO Box 117, Rockford, IL 61105, USA.

Pierce & Warriner, 44 Upper Northgate Street, Chester, CH1 4EF, UK.

Promega

Promega Ltd, Delta House, Enterprise Road, Chilworth Research Centre, Southampton, UK.

Promega Corporation, 2800 Woods Hollow Road, Madison, WI 53711–5399, USA.

Qiagen

Qiagen Inc., c/o Hybaid, 111–113 Waldegrave Road, Teddington, Middlesex, TW11 8LL, UK.

Qiagen Inc., 9259 Eton Avenue, Chatsworth, CA 91311, USA.

Rockland Technologies

Rockland Technologies Inc., PO Box 316, Gilbertsville, PA 19525, USA.

c/o *Oncogene Sciences SA*, 18 Rue Goubet, 75019 Paris, France.

Rohm

Rohm Pharma GmbH, Kirschenalle, Darmstadt, D-64293, Germany.

Rohm Tech. Inc, 195 Canal Street, Malden, MA 02148 USA.

Sartorius

Sartorius Corporation, 131 Edgwood Boulevard, Edgewood, NY 11717, USA.

Sartorius AG, Weender Laudstrasse 94-108, D-37075 Goettingen, Germany.

Schleicher and Schuell

Schleicher and Schuell Inc., Keene, NH 03431A, USA.

Schleicher and Schuell Inc., D-3354 Dassel, Germany. Schleicher and Schuell Inc., c/o Andermann and Company Ltd.

Schott

Schott Glass, Schott Glasswerke, Hattenbergstrasse 10, Mainz, D-55122, Gemany. *Schott Corporation*, Schott Parenta Systems, 3 Odell Place, Yonkers, NY 10701, USA.

Seikagaka Kogyo

US distributors: 30 West Gude Drive, Suite 260, Rockville, MD 20850-1161, USA.

UK distributors: ICN Biochemicals Ltd, Eagle House, Peregrine Business Park, Gomm Road, High Wycombe, Bucks. HP13 7DL, UK.

Sepracor, *see* Biosepra.

Shandon Scientific Ltd, Chadwick Road, Astmoor, Runcorn, Cheshire WA7 1PR, UK.

Sigma Chemical Company

Sigma Chemical Company (UK), Fancy Road, Poole, Dorset BH17 7NH, UK.

Sigma Chemical Company, 3050 Spruce Street, PO Box 14508, St. Louis, MO 63178–9916, USA.

Silverson, Machines Ltd, Waterside, Chesham, Bucks. HP5 1PQ, UK.

Sorvall DuPont Company, Biotechnology Division, PO Box 80022, Wilmington, DE 19880–0022, USA.

Sterling Organics, Hadrian House, Favdon, Newcastle-upon-Tyne NE3 3TT, UK; 33 Riverside Avenue, Reusselaer, NY 12144, USA.

Stratagene

Stratagene Ltd, Unit 140, Cambridge Innovation Centre, Milton Road, Cambridge CB4 4FG, UK.

Strategene Inc., 11011 North Torrey Pines Road, La Jolla, CA 92037, USA.

TSK TosoHaas, 156 Keystone Drive, Philadelphia, PA 18936, USA; 7 Lonsdale, Linton, Cambs. CB1 6LT, UK.

Unisyn Technologies Inc., 25 South Street, Hopkinton, MA 01748, USA.

United States Biochemical, PO Box 22400, Cleveland, OH 44122, USA.

Upstate Biotechnology

Upstate Biotechnology Inc., 199 Saranac Avenue, Lake Placid, NY 12946, USA.

c/o *TCS Biologicals Ltd*, Botolph Claydon, Bucks. MK18 2LR, UK.

Varian

Varian Ltd (UK), 28 Manor Road, Walton-on-Thames, Surrey KT12 2QF, UK.

Varian Associates, Dept 87, PO Box 9000, San Fernando, CA 91340, USA.

V-Labs Inc., 423 N. Theard St., Covington, LA 70433, USA.

Wellcome Reagents, Langley Court, Beckenham, Kent, BR3 3BS, UK.
Whatman
Whatman International Ltd, Whatman House, St Leonards Road, Maidstone, Kent ME16 0LS, UK.
Whatman Inc., 6 Just Road, Fairfield, NJ 07004, USA.
Qiagen GmbH, Max volumer strasse 4. 40724, Hilden, Germany.

Index

All reference to figures are given in italics.

Index

Index